Fossil Plants

Dec 2009

To Gladys

Greetings from

Eva Koppelus
& [signature]

IDENTIFICATION GUIDE TO THE

Fossil Plants

OF THE HORSESHOE CANYON FORMATION OF DRUMHELLER, ALBERTA

Kevin R. Aulenback

UNIVERSITY OF
CALGARY
PRESS

University of Calgary Press
2500 University Drive NW
Calgary, Alberta
Canada T2N 1N4
www.uofcpress.com

LIBRARY AND ARCHIVES CANADA CATALOGUING IN PUBLICATION

Aulenback, Kevin R., 1960-
   Identification guide to the fossil plants of the Horseshoe Canyon
Formation of Drumheller, Alberta / Kevin R. Aulenback.

Includes bibliographical references and index.
ISBN 978-1-55238-247-9

   1. Plants, Fossil–Alberta–Horseshoe Canyon Formation–Identification.
2. Paleobotany–Alberta–Horseshoe Canyon Formation–Guidebooks.
3. Palynology–Alberta–Horseshoe Canyon Formation–Guidebooks.
4. Paleobotany–Cretaceous. I. Title.

QE938.A4A94 2009   561'.1971233   C2009-902711-9

The University of Calgary Press acknowledges the support of the Alberta Foundation for the Arts
for our publications. We acknowledge the financial support of the Government of Canada through
the Book Publishing Industry Development Program (BPIDP) for our publishing activities. We
acknowledge the financial support of the Canada Council for the Arts for our publishing program.

We thank the Alberta Historical Resources foundation for its support of this project.

Printed and bound in Canada by Marquis
This book is printed on FSC Flo Dull paper

Cover design by Melina Cusano
Page design and typesetting by Melina Cusano

*This book is dedicated to
my wife, Julie,
and my children, Brandy, Spencer and Lucas,
who have given me invaluable understanding and support.*

*It is also dedicated to those researchers with whom
I have collaborated throughout my professional career
and especially the late Dr. Beth McIver.*

Contents

Preface

This guide is designed to aid the professional and amateur alike in the identification of known fossil plants in the Horseshoe Canyon Formation of Drumheller, Alberta, Canada. This is not a scientific textbook, although it is assumed that the reader will know or be able to understand some of the terminology used in the scientific description of plants. A glossary is included with definitions.

It is well known that many plants in the fossil record are only represented by a few parts or organs. With some of the more well-known plants, photographs or drawings are given to help aid in the identification of individual plant parts that may be encountered. Some plant identifications are controversial and these are elaborated upon with their identifications.

Many as yet unidentified plants are also included with the hopes that someone may have encountered the plant, or plant part, in a fossil or living flora elsewhere and will aid in its identification.

The environmental setting under which these plants grew in the formation is discussed.

Foreword

by Philip J. Currie
Professor, University of Alberta
and former Curator of Dinosaurs,
Royal Tyrrell Museum of Palaeontology

I have always liked fossil plants. In fact, I sometimes have thought that if I had not decided to become a specialist in fossil vertebrates, then I would have pursued a research career as a palaeobotanist. Whereas my interest in dinosaurs went back to the tender age of six, I became interested in plant fossils through lectures in undergraduate courses at the University of Toronto. When I started my career as a vertebrate palaeontologist at the Provincial Museum of Alberta (now the Royal Alberta Museum), I enrolled as a special student at the University of Alberta so that I could take the palaeobotany course offered by Dr. Wilson Stewart. An excellent scientist and entertaining lecturer, Dr. Stewart enhanced my interest in the subject. Coincidentally, my wife (Dr. Eva Koppelhus) is a specialist in palynology, which is a subsection of palaeobotany centred on fossil pollen and spores. Like any pair of professionals, we sometimes engage in good-natured jibes about the relative merits of our specializations. Ultimately though, Eva has the trump card up her sleeve – without plants there would have been no dinosaurs!

Kevin Aulenback was a young and enthusiastic student studying at the Northern Alberta Institute of Technology to be a laboratory technician when we hired him to collect and prepare dinosaurs in 1981. There was no doubt about his talents and potential, so he was asked to move to Drumheller when we started the Tyrrell Museum of Palaeontology the following year. Those were crazy and rather stressful times when Kevin and the rest of the team set out to build a world-class facility under a schedule that most of our colleagues thought was impossible. He was soon assigned his own team of workers to collect and prepare dinosaurs and other types of fossils. When the museum finally opened in 1985, all of us knew that it was one of the greatest experiences and opportunities in our young lives.

In spite of the fact that he had a fulltime job working with fossils, Kevin would also use his own time to push the boundaries of his chosen profession. In 1984, he went out and found (within the city limits of Drumheller) the first mammal jaw from the Cretaceous rocks of the Horseshoe Canyon Formation. Almost on a whim, a few years later he threw some plant-laced rock from behind the Tyrrell Museum into acid (Aulenback and Braman, 1991) to see what was inside. Much to everyone's surprise, the result was exquisitely preserved, three-dimensional plant fossils that earned him his first appearance in *National Geographic* (Gore 1993, Psihoyos 1994) and his first scientific paper (McIver and Aulenback, 1993). Ultimately, it also led to the book that you are now holding.

Kevin has always been great in both the laboratory – where he is very skilled in the preparation of small specimens, although equally comfortable working on large fossils – and in the field. Some could say that he has been very lucky with some of his discoveries. However, "luck" is a combination of being the right person in the right place (usually just at the right time), coupled with well-honed powers of observation. In 1987, we were working together looking for dinosaur eggs in the Milk River area of southern Alberta. When he did not show up on time on the last day of the expedition, I knew that something was up. Sure enough, when he finally appeared, he was so excited that it was difficult to follow the barrage of words. But the words "eggs, babies, and embryos" were all that was necessary to inspire me to follow him back the way he had come. He led me to a hillside covered by eggshell fragments, mixed with the miniature bones of duck-billed dinosaurs! Without a doubt, one of the most exciting discoveries that I have ever been involved in, it is no surprise that we were like a couple of kids as we picked up the bones and fragments of eggs that had cascaded down the slope. After dropping a tooth-filled jaw of an embryonic hadrosaur (because my eyes and brain could not believe what they were seeing), we decided to pull out of Devil's Coulee long enough to pick up the necessary tools and supplies from the museum. Little did we know at the time, but it would be many months before any of us would see the site again (for fuller accounts of Kevin's exciting discovery, see Currie [1988], Grady [1993], and Acorn [2007]).

The following year in China (Grady 1993), Kevin returned from the field almost daily with the best discoveries. One day, tired of being bested by the best, I told him he had to walk back to camp behind me. As we set off down the trail through the badlands of the Gobi Desert, I felt confident that I would make the big find of the day. And finally, at a fork in the trail, I saw the characteristic glint of dinosaur eggshell along the path to the right. It only took a few seconds to confirm that my discovery was not significant. However, that was all the time that was necessary for Kevin to surge ahead of me on the left trail. Within seconds, he called us over to see a beautiful circle of oviraptorid ("egg thief") dinosaur eggs. If the situation had not been so funny, I probably would have got angry with him.

In 1990, Kevin wanted to broaden his preparation experiences beyond dinosaurs (which tend to be rather large) by working on smaller, more manageable specimens. He is extremely good at carefully but quickly preparing the rock away from small delicate specimens, so it is no surprise that the fossil plants he was preparing soon led to his second scientific publication (Aulenback and Lepage 1998). At this time,

Kevin also became highly proficient at the preparation of palynological samples. His skills would later help other labs to find the K/T fern spike in Alberta.

After the discovery in 1996 of the first feathered dinosaurs of Liaoning, I went to China often to work on some of the most spectacular specimens that had ever been discovered in the twentieth century. When the opportunity arose to take a technician with me to China to prepare some of the specimens, Kevin was the first name that came to mind. I had to pull some strings to get him back into working on dinosaurs, but it eventually worked. We met in Beijing on a frosty December morning in 1997, and shortly after were huddled over a table of specimens in the National Geological Museum of China. As the work progressed under difficult, poorly lit conditions, it became evident that two of the specimens were not the "feathered" dinosaur species (*Protarchaeopteryx robusta*) that they had been identified as. The legs of this turkey-sized animal were relatively longer, and the arms were relatively shorter. The long days working on the specimens and fruitful evenings of discussion (when we weren't playing darts in the lounge) back in the hotel eventually led to the description of *Caudipteryx zhoui*. The scientific paper in *Nature* (plus a companion popular article in *National Geographic* [Currie 1998]) of this incredible little dinosaur was pivotal in the fiercely raging controversy about whether or not birds had originated from dinosaurs. In a single animal we found a mosaic of characters that were not supposed to be found in early birds if one assumed that they had originated from any group of animals other than dinosaurs. In addition, however, *Caudipteryx* also had feathers behind the arms and at the end of the tail that are indistinguishable from those of modern birds. Many additional species of feathered dinosaurs have been found since the original publications appeared on *Caudipteryx*, but none swayed the argument to the same degree in favour of the origin of birds from dinosaurs.

Back in Alberta, Kevin began a five-year project, the totally revamping of the palaeoconservatory at the Tyrrell Museum as researcher, designer, and installer. This remarkable display of living plants is strategically placed to lessen the effects of what museologists refer to as "museum fatigue." But as Kevin saw it, the aesthetically pleasing oasis in the middle of Alberta's badlands also had to be educational. Using his knowledge of Cretaceous fossil plants and their living relatives, he assembled a remarkable display of plants that resemble the species that lived in the region some 75 million years ago. It is visually pleasing enough to make the palaeoconservatory a favourite place for wedding photos, but also serves as a source of information for palaeontologists who want to conceptually reassemble their bits and pieces of fossilized roots, leaves, stems, flowers, and seeds into full plants. I am sure it was while Kevin was digging, clipping, and watering that many of the ideas for this remarkable book developed!

Talents come in many forms and are gauged by different standards in different fields of endeavour. Kevin is one of those rare people whose industry and enthusiasm, combined with acute powers of observation and boundless curiosity, inspires the description of "talented" in anything he gets interested in. It can be technical or intellectual, dead or alive, plants or animals, descriptive or creative – you name it. He can look proudly on many accomplishments, including beautifully prepared and mounted specimens on display in the Royal Tyrrell Museum of Palaeontology (Drumheller) and the Institute of Vertebrate Paleontology and Paleoanthropology (Beijing), media coverage that includes multiple mentions in *National Geographic*, scientific publications, published photographs, artwork, and a delightful family that includes three children. This volume represents another phase in a remarkable career that is probably already heading off in new directions!

References

Acorn, J. 2007. *Deep Alberta: Fossil Facts and Dinosaur Digs*. Edmonton: University of Alberta Press.

Aulenback, K.R., and D.R. Braman. 1991. A chemical extraction technique for the recovery of silicified plant remains from ironstones. *Review of Palaeobotany and Palynology* 70:3–8.

Currie, P.J. 1988. The discovery of dinosaur eggs at Devil's Coulee. *Alberta* 1:3–10.

Currie, P.J. 1998. Caudipteryx revealed. *National Geographic* (January): 86–89.

Gore, R. 1993. Dinosaurs. *National Geographic* (January): 2–53.

Grady, W. 1993. *The Dinosaur Project*. Toronto: Macfarlane, Walter and Ross.

McIver, E.E., and K.R. Aulenback. 1993. Morphology and relationships of Mesocyparis umbonata sp. nov.: Fossil Cupressaceae from the Late Cretaceous of Alberta, Canada. *Canadian Journal of Botany* 72: 273–95.

Psihoyos, L. (with H. Knoebber). 1994. *Hunting Dinosaurs*. New York: Random House.

Acknowledgments

I extend my appreciation to: Dr. Arthur Sweet of the Geological Survey of Canada, Calgary, Alberta, for his years of support and encouragement in both the fields of Palynology and Palaeobotany; Technician Jenny Woo, who, over the years, S.E.M.-scanned most of the images shown; Dr. Steven Manchester, who suggested this book and gave constructive criticism for the angiosperm section; Dr. James Basinger and the late Dr. Beth McIver from the University of Saskatchewan for introduction to the field of palaeobotany and discussions over the years on conifers; Dr. Dennis Braman of the Royal Tyrrell Museum for editing an earlier version of this book and allowing me the latitude to discover plant fossils no matter how small during my years as a professional technician; Marilyn Laframboise for her literary help with the book and her development of an index; Dr. Eva Koppelhus, who S.E.M.-scanned images of the Marsileaceae; Dr. Peter Crane for discussions on fossil plants over the years; Dr. Diane Erwin and Dr. Kathleen Pigg for early discussions on monocots; Kent Wallis, Wayne Marshall, Donna Sloan, and Maurice Stefanuk for showing plant sites to me in the formation over the years; Wendy Sloboda for showing seed sites to me in the Oldman Formation and Darren Tanke for showing a rhizome site to me in Dinosaur Provincial Park; Dr. Phillip Currie of the University of Alberta for introducing me to the field of palaeontology many years ago; my wife, Julie Aulenback, and my children Brandy, Spencer and Lucas, for their support and encouragement and lastly to all my other friends too numerous to mention for their support.

For funding contributions, I wish to thank the Dinosaur Research Institute, Calgary, Alberta, Bill and Karen Spencer, William and Irene Bell, Greg Pavan and family, Ken and Gina Mah, Dougald (Duke) and Lucielle Aulenback, Danny Mah and family, Dougald (Doug) Jr. Aulenback and family, Alfred (Fred) Orosz, Karen Lai-wan Leung, Steven Manchester, James F. Basinger, Ben A. LePage, and Michael Ryan. A special thanks to Al Rasmuson and Anita Fryters for help in getting funding organized.

Photographic Credits

All photographs are courtesy of the author, except figures 8–13, 21, 38–40, 47, 55, 65, 152, 163, 185, 193, 197, 207, 208, 213–215, 228, 270, 283, 355, 437, 479, 498, 527, 549, 550, 554, 562–564, 566, 581, 583, 589, 591, 594–603, and 710, courtesy of D. Braman, Royal Tyrrell Museum of Paleontology; figure 806, courtesy of D. Brinkman, Royal Tyrrell Museum of Paleontology; figures 51, 102–104, 252, 499, and 518, courtesy of A. Sweet, Geological Survey of Canada. R. Otsian for allowing the photography of specimen figure 247 from her collection.

Figures 242, 424, 429, 431, 434, 435, 438–442, 444–446, 448, 463, 467–473, and 692–695 are reproduced with permission from the *Canadian Journal of Botany* 72 (1994), and 83 (2005); figures 304, 305, 379, 578, 664, and 665 are reproduced with permission from the *Review of Palaeobotany and Palynology* special issue 70, 1/2 (1991); figures 343–347, 349–351, 354, 356–359, 362–369, 372, 373, and 375 are reproduced with permission from the *International Journal of Plant Sciences* 159,(1998).

Figure 546 is drawn from a photograph with the permission of S. Manchester. Figures 125 and 127 are drawn from photographs with the permission of R. Weber. Figures 573, 574, 580, 587, 592 are drawn from photographs with permission from R. Serbet.

The sorus in figure 68 is reproduced with permission from A. S. Foster and E. M. Gifford, Jr., *Comparative Morphology of Vascular Plants* (San Francisco: W.H. Freeman, 1974).

Special thanks to the Royal Tyrrell Museum of Palaeontology and the University of Alberta for allowing photography of specimens in their collections.

INTRODUCTION

Many of us have wondered at the great diversity of plants we see all around us. Where do they come from? How are they related? Many of the answers can be found through the study of paleobotany, the study of ancient plants. Plants have been evolving since the Precambrian, approximately 550 million years ago. This book has a very narrow but important focus. It looks at a small segment of their evolution when flowering plants were dominating the ancient landscape. The time is the Upper Cretaceous.

Although many different floral provinces existed worldwide, here we home in on a section of land, the small beachfront property of the delta plain in the northern hemisphere, the Horseshoe Canyon Formation in the vicinity of Drumheller, Alberta, Canada. It is here we find the mighty giants, the dinosaurs.

Most people first encounter the land of the dinosaurs through visits to their museums. In many museums, the dinosaurs, although artistically done, are usually skeletons placed in dynamic poses with the surrounding area a stark contrast of metal or sand. There is no hint of the great floral diversity present where they once lived. The flora supported their growth, fed them, and hid them. Without plants there would not have been any dinosaurs or, for that matter, man.

The Upper Cretaceous Horseshoe Canyon Formation was deposited at a time of great floral overture. The gymnosperms of past ages were already under pressure from the rapidly evolving faster-growing angiosperms. Wet/dry seasonality was prominent with a maritime influence from the inland sea. Our latitude was even more northerly than that of today with Drumheller at approximately 56° N latitude 72–73 million years ago at the beginning of the formation's deposition and even further north at approximately 60°15" N latitude after deposition in the early Paleocene. Presently the Drumheller area sits at 51°15" N latitude 112°24" W longitude and is a mid-continental, dry, cool, temperate climate.

It is hoped this book will give the reader a better understanding of the great floral diversity that existed at this well-known place and time in history.

PURPOSE

In 1949 Dr. W.A. Bell of the Geological Survey of Canada published the "Uppermost Cretaceous and Paleocene Floras of Western Canada," Bulletin #13. To date the publication has been the most complete source of information available on the Horseshoe Canyon Flora. This bulletin is used today by researchers and laypersons alike as an aid and guide in the identification of fossil plants. Unfortunately, this scientific publication is greatly outdated and much has been discovered since its publication.

The present identification guide has many purposes. The first is to give a much-needed summary of information and new ideas on the fossils that occur in the formation. Many scientific finds have been made over the ensuing years, but the reports are scattered throughout various journals and publications. The guide also contains many unpublished fossil finds noted by the present author over sixteen years of discovery, observations, and interpretations of paleobotany in the Horseshoe Canyon Formation.

It is hoped this guide will give the amateur collector as well as the professional a starting point for the identification of their finds in this and other similar formations. Although researchers in the field do find many fossils, amateurs find important specimens as well.

Examples include the site called "Kent's knoll" and specimens such as *Taxodium wallisii* named after Mr. Kent Wallis of Drumheller, who led me to the site, as well as *Wessiea oroszii* named after Mr. Alfred (Fred) Orosz of Nacmine, who also led me to another site. An extinct new taxon of the Cycadophyta is being described based primarily on a single specimen found by a young man, Mr. Alex Bramm, who donated the excellently preserved seed cone to the Royal Tyrrell Museum of Palaeontology. The specimen appears pivotal in re-describing structural morphology of other fossil cones found around the world. Landowners also deserve credit, including Mrs. Lowen, who graciously allowed access to her property to collect additional specimens of these cones for morphological studies. There are those who merely pick up a rock of interest such as an inquisitive young boy who donated a fossil without leaving his name. Fossils like *Albertarum pueri*, whose species name means "male child," are very important to the science of paleobotany and shed light on the evolution of plants. The role of the amateur collector cannot be overstressed, but it is only through proper documentation and scientific study by professional paleobotanists that these finds can shed light on the history of plant life.

This guide is fairly specific to the Horseshoe Canyon Formation, although the information on the fossil plants may be used as a guide to identify other plant forms in other formations. Where possible, scientific names are included. The word "-like" is used when the plant identification as to genus cannot be scientifically confirmed, such as *Taiwania*-like. All measurements are given in metric.

This guide is not a scientific journal, but it assumes you know a little about botany. It discusses the different ways in which the Horseshoe Canyon Formation is dated and divided into finer time segments. It describes types of preservation found in the formation and provides a complete history of fossil plant collecting in the area.

Paleobotany encompasses both the sciences of paleobotany (ancient plants) and paleopalynology (ancient pollen and spores). Both are important disciplines in the field of botany (living plants). Whenever possible, this book gives examples of both plants and palynomorphs.

Since the book is tailored to plant remains that can be seen with the unaided eye in the field or with a dissecting microscope, a history of the palynomorphs will not be attempted. This is not to say that they are unimportant; the history of the palynomorphs is extremely diverse and varied but would need and deserves to be done as a separate investigation. The plants and palynomorphs are used together to help give a clearer picture of the plant diversity of the time. Comparisons with living floras are also given.

Some controversial discussions on various fossil plant finds in the formation are included to spark further thought and debate by others. Interpretations given of these finds and others are unpublished but can all be substantiated by previously written scientific papers or the fossils themselves. These interpretations are based on logical assumptions and are solely those of the author. Acceptance or rejection of any interpretation given is at the discretion of the reader.

Although this is a guide for use in the identification of fossil plants, it contains many photos of living plants for comparison or example. In the formation many plants do not have identifiable fruit or leaf remains and are identified based solely on palynomorphs. In such cases living plant examples are used. It should be emphasized that a fossil form, if found, may not resemble its living counterpart and that the living leaf forms are merely representatives of what may be encountered. Also, many living families and genera can have highly variable leaf shapes from species to species or in a single species.

GEOLOGY OF THE HORSESHOE CANYON FORMATION

Fig. 1. Map of Red Deer River section and surrounding towns.

The Horseshoe Canyon Formation consists of approximately 100 kilometres of exposure along the Red Deer River valley and tributaries from the town of Dorothy in the south to the town of Ardley in the north (Fig. 1). The formation forms part of the Edmonton Group, which also includes the overlying Whitemud, Battle, and Scollard formations. In subsurface the Edmonton Group forms an arc through the southern region of the province (Fig. 2). The Horseshoe Canyon Formation is underlain by the marine Bearpaw Formation with the Whitemud Formation capping its upper limit (Fig. 14).

The Horseshoe Canyon Formation spans parts of two stages, the Campanian and the Maastrichtian of the Upper Cretaceous. Coal seam 10 and below are in the Campanian while the remainder of the formation, above coal seam 10, is in the Maastrichtian.

The formation has been subdivided in many different ways. Some of the most commonly accepted ways are listed below.

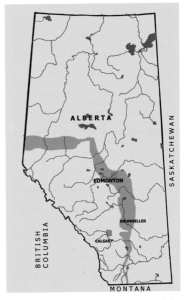

Fig. 2. Map of subsurface extent of the Edmonton Group (Horseshoe Canyon Formation) in Alberta.

Ash Beds

Although there are numerous ash beds in the formation, only two have been dated (Fig. 14). These ash beds are dated using potassium/argon dating techniques. This is the measure of the accumulation of argon from decomposition of potassium in a mineral. This clock is usually set in the geological record by heat in the form of volcanic eruptions.

Potassium/argon dating uses the mineral biotite. This is a crystal formed during volcanic eruptions and is usually abundant in ash. Potassium/argon dating has been used to date rocks up to 4 billion years old unlike carbon 14, which can only date to 40–50,000 years. Potassium/argon dating has a margin of error anywhere from 0.25 to 2 million years. Although seemingly long, it is short in an overall geological context.

The presently dated ash beds in the formation are from immediately below coal seam 10 at 70.6 mybp (million years before present) and in coal seam 11 at 67.0 mybp. These are bracketed by two ashes, one from below the Horseshoe Canyon Formation in the upper Bearpaw Formation at 73.2 mybp and one from above the Horseshoe Canyon Formation in the Battle Formation at 66.8 mybp. Based on these dates, the Horseshoe Canyon Formation spans from about 66.9–72.5 mybp or approximately 5.6 million years of time.

Coal Seam or Zone Markers

The Horseshoe Canyon Formation also contains thirteen coal seams or zones that can be used as stratigraphic markers (Fig. 14). These coals, which in past literature were identified as seams (Gibson 1977), can sometimes bifurcate or split laterally and can form coal sets. These sets and individual seams are now referred to as *coal zones*. Coal zones can be traced for several tens of kilometres laterally and because of their overlap are very useful in stratigraphy.

Coal zone 0 is a single seam and is exposed from East Coulee (Fig. 3) to the Hoodoos. This is a thin seam and has never been commercially mined. Mega-fossils have not been reported from this seam.

Coal zone 1, also called the Drumheller seam, is a single seam that extends in exposure from East Coulee to the swinging bridge, just opposite the junction of highway 56, and Cambria. This seam ranges from 0.9–3.3 m thick. This seam has been commercially mined in the past and many large stumps and logs have been excavated from it. In the exposed coal, only a few rare stumps can now be found.

Coal zone 2 is above coal zone 1 and is exposed from the town of East Coulee to just south of the Drumheller Cemetery. This seam bifurcates at Willow Creek (Fig. 4) and northward forming two closely associated seams. The seam varies from

0.2–1.8 m in thickness and has been commercially mined in the past. In the hills just above the town of East Coulee, a fossil forest of *in-situ* stumps can be seen in the exposed coal. Fossilized stumps in the coal can also be seen in the area of the Hoodoos.

Coal zone 3 extends from East Coulee to just south of the town of Drumheller (behind the composite high school). This seam ranges in thickness from 0.2–0.7 m. Just above the town of East Coulee, the seam contains numerous fossil logs but fossil stumps are rare.

Coal zone 4 extends from East Coulee to Michichi Canyon, just north of the town of Drumheller. This coal also bifurcates into two and ranges from 0.2–0.5 m in thickness. Only small amounts of fossilized wood have been found in this seam with the odd *in-situ* tree stump.

Coal zone 5, also called the Newcastle Seam, extends from south of the Wade Ranch (just east of the Cambria Bridge) to the town of Drumheller. This is a single seam ranging from 0.3–1.6 m in thickness. Coal Seam 5 was commercially mined in the past. In the Willow Creek area, a small fossil forest of *in-situ* stumps can be seen in the coal.

Coal zone 6 is a single seam exposed from Rosebud Creek to the Drumheller Golf Course. The seam ranges from 0.2–0.5 m thick. Small tree stumps and isolated logs have been found in this seam around Drumheller.

Coal zone 7, also called the Daly Seam, is exposed from the town of Wayne to north of the Drumheller Golf Course and is viewable outside the Royal Tyrrell Museum (Fig. 5). This seam bifurcates into multiple smaller seams laterally and vertically. This seam ranges from 0.1–3.3 m in thickness. Small isolated fossil stumps and logs

Fig. 3.   View of East Coulee area coal zones 0–3.

Fig. 4.   View of Willow Creek area coal zones 2–5.

Fig. 5.   View of Royal Tyrrell Museum area coal zones 6–8.

can be found. This seam was commercially mined for a limited time in the past.

Coal zone 8 is more of a coal set as it is sometimes exposed as three closely spaced seams. These seams are exposed from just north of the town of Drumheller on the west side of the valley to just north of the Morrin Bridge crossing. The seam also extends south to the town of Wayne. The total seam thickness ranges from 0.1–2.0 m. It did not contain sufficient coal for mining.

Coal zone 9 runs from Drumheller to about six kilometres north of the Morrin Bridge. It also is more of a coal set and actually disappears in some localities. This seam normally ranges from 0.1–0.7 m in thickness. Due to its lack of thickness, this seam has received very little mining attention.

Fig. 6.   View of Horsethief Canyon area coal seams 8–10.

Fig. 7.   View of Horseshoe Canyon coal zones 10–12.

Coal zone 10, also called the Marker Seam, is exposed at Horsethief Canyon (Fig. 6) in the Drumheller Marine Tongue up river to the mouth of Big Valley Creek. This seam is also exposed in Horseshoe Canyon. This coal is very thin from 0.01 to 0.03 m thick. Although it contains little coal, it is generally the most conspicuous seam in the valley. This seam has not been mined.

Coal zone 11, the Carbon seam, is exposed upstream of the Morrin Bridge up to the river valley east of Lousana. It also has excellent exposure in Horseshoe Canyon (Fig. 7). Silicified wood fragments have been found in this coal. This seam is from 0.2 to 1.6 m in thickness and has been mined in various areas in the past.

Coal zone 12 or the Thompson seam is exposed from north of the Morrin Bridge crossing to east of Lousana. The seam is also exposed in Horseshoe Canyon. The Thompson seam varies from 0.1 to 1.5 metres in thickness. The coal was mined in one area only for a brief period and is no longer being mined.

Palynology

Palynology is the study of recent or fossil acid-resistant microfossils, including pollen and spores. These spore and pollen (palynomorphs) are, due to their resistant nature, all that sometimes remains of a fossil plant.

The formation has been scientifically divided into palynological (spore and pollen) zones (Fig. 14). Each zone represents a time period characterized by pollen or spores, and boundaries are determined by first appearance or last occurrences of species. Many spores and pollens may enter or depart at each zone, although the zone is named after one pollen or spore type.

There are six palynomorph zones defined in the Horseshoe Canyon Formation (Srivastava 1970). These stratigraphic markers are used to plot important fossil discoveries in the formation.

Zone 1 is the transition zone and marks the departure of the Bearpaw Sea. This area has not produced recognizable plant fossils to date. It is approximately 28.8 m thick and is topped by 60 cm of shale.

Zone 2 or the *Aquilapollenites leucocephalus* (Fig. 8) zone is approximately 26.2 m thick and ends at the basal shale of coal seam number 4. There are many plant fossils known from this area.

Zone 3 is the *Wodehouseia edmonticola* (Fig. 9) and *Wodehouseia gracile* zone. This zone starts at the base of the number 4 coal seam and ends at the base of coal seam number 7 and is approximately 18.3 m thick. This zone has the potential to produce large quantities of plant fossils not only from the coals but also from shale and ironstones. Preliminary investigations have been very encouraging.

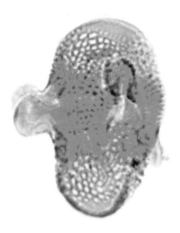

Fig. 8. *Aquilapollenites leucocephalus* pollen.

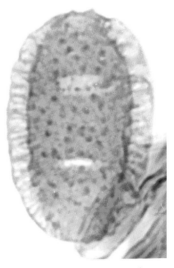

Fig. 9. *Wodehouseia edmonticola* pollen.

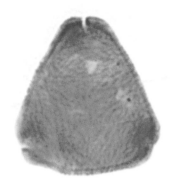

Fig. 10. *Pulcheripollenites krempii*
pollen.

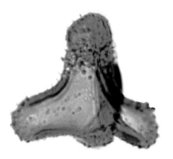

Fig. 11. *Mancicorpus borealis* pollen.

Fig. 12. *Mancicorpus vancampoi* pollen.

Zone 4 or the *Pulcheripollenites krempii* (Fig. 10) zone has produced the most varied plant floras from a variety of sites. This zone is about 64.0 m thick. It starts at the base of coal seam number 7 or Daly seam (3.2 km west of little church below road level) and ends at the top of a massive sandstone layer above coal seam number 9.

A sub-zone is placed at the top of zone 4, called the *Mancicorpus borealis* (Fig. 11) sub-zone. This sub-zone is short and extends vertically for only 7.9 m, ending at the top of a shale bed underlying a 1.8 m thick sandstone. Very little is known about plant fossils in this interval.

Zone 5 or the *Mancicorpus vancampoi* (Fig. 12) zone is approximately 44.5 m thick. This zone takes in the Drumheller Marine Tongue. This zone does not include coal seam 11, the Carbon seam, but its top is marked by increased organic deposits. This zone also has the ability to produce recognizable plants fossils but investigations to date have been sparse.

Zone 6 is the *Scollardia trapaformis* (Fig. 13) zone. This zone extends to 1.5 m above the number 12 or Thompson coal seam and is approximately 10.0 m thick. Very little is known about plant mega-fossils from this zone.

Fig. 13. *Scollardia trapaformis* pollen.

Palaeo-Magnetostratigraphy

This method gives as great a resolution as palynology and when used together with it can give even greater stratigraphic resolution. Palaeo-magnetostratigraphy relies on shifts in the Earth's magnetic poles. As sediments are laid down, dissolved iron minerals align magnetically like the needle of a compass. These magnetically aligned particles are preserved in the sediments and can be read.

A series of letter/numbers were developed to represent various ages. The "C" series system starts with the present as C1 normal or "C1n." The C series ends at C34n, which spans from the Campanian/Santonian boundary to the early Aptian (Cretaceous). The preceding letter "C" is often not used in this series.

It is known that over the history of the earth that many polarity reversals have occurred at irregular intervals. These reversals are given the letter "r" after their number. When grains are found that align with the present magnetic polarity they are designated with a number followed by the letter "n" for normal. Thusly any polarity number can be either normal or reversed ie. 30n or 30r. With continuing studies these basic numbers have been divided into more precise subdivisions ie. 32n. 3r or in 32 normal polarity 3rd reversal polarity. These Numbers or Paleo chrons can be correlated across many different, yet contemporaneous, formations. For more details see Ogg (1995).

There are up to seventeen palaeomagnetic zones in the formation, which span from 32r.1r to 30r (Fig. 14) (Lerbekmo and Coulter 1984, Lerbekmo and Braman 2002).

Fig. 14. Stratigraphic column showing the different ways in which the formation has been divided. W/B = Whitemud and Battle formations.

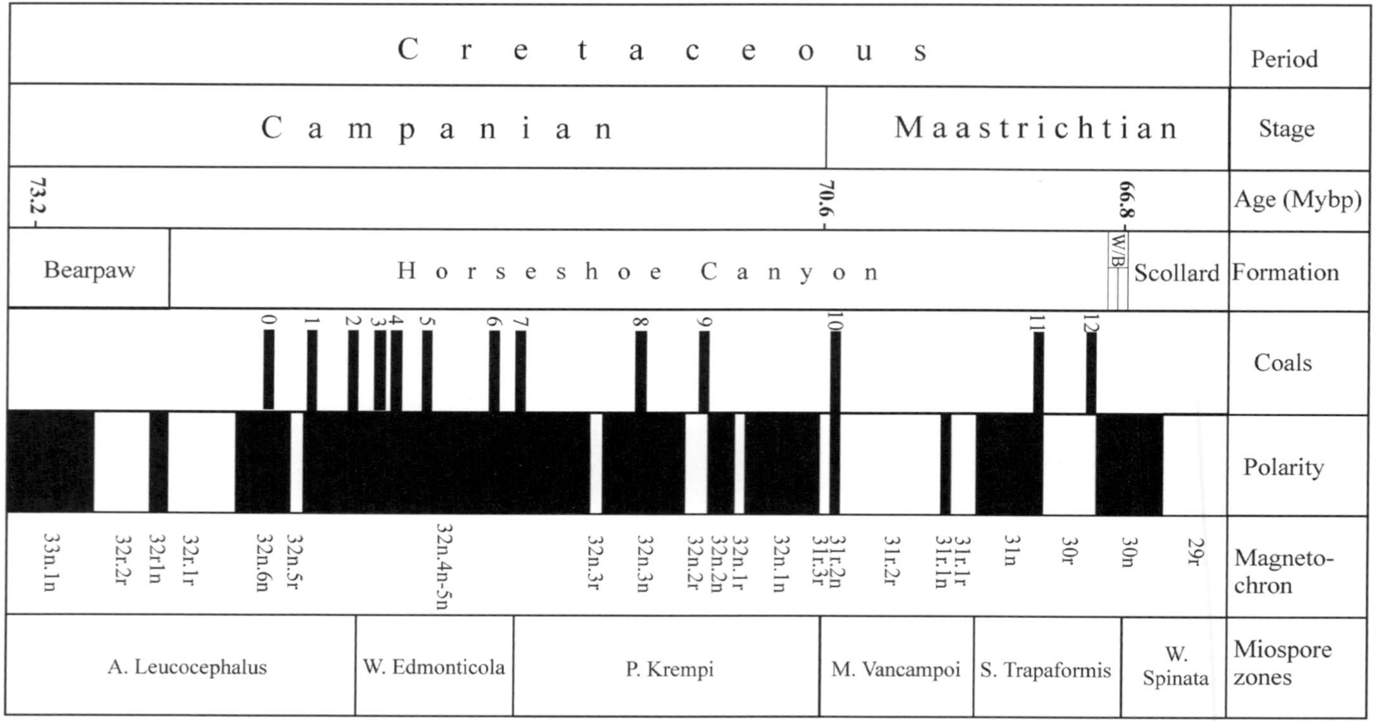

TYPES OF PRESERVATION AND METHODS OF STUDY

The type of preservation of a fossil plant dictates how it can be studied and what techniques should or can be used to obtain information. Because of this, it is very important to understand the preservational mode of the fossil.

In the Horseshoe Canyon Formation, plant fossils are preserved as compression impressions, casts and moulds, petrifications, and carbonized remains.

Compression Impressions

Many plant fossils are preserved as compressions or impressions. Leaf beds contain leaves that were trapped in silts and fossilized. Upon splitting rocks these leaves are exposed along the freshly broken surface or laminae (Fig. 15). Seed cones can also be found this way but show a cross-section along the fracture.

Casts and Moulds

A cast is the internal imprint of a plant part. Plants may have hollow stems or contain soft internal tissue that can decay or desiccate quickly and leave internal hollows. These hollows can in-fill with sediment and fossilize.

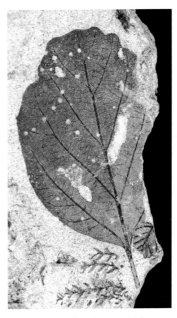

Fig. 15.   Fossil angiosperm leaf compression.

Fig. 16. *Parataxodium* sp., natural mould of seed cone.

Fig. 17. *Cunninghamia* sp., natural mould of branch.

During erosion these internal "structures" may roll free of the surrounding matrix. Surface detail on these structures may reveal internal or external cell structure, depending upon whether or not the surrounding external plant tissue is incorporated or adheres to the cast structure.

Natural moulds can be found in a wide variety of fossils from vertebrates and invertebrates to plants. Moulds occur when the fossil weathers faster than the surrounding matrix (rock) (Figs. 16, 17).

In the Drumheller area, plants are commonly preserved as carbon, organic trace or calcite replaced, all of which erode quite quickly upon exposure to the elements. If the fossil is entombed in a resistant sandstone or mudstone (ironstone), a natural mould may be left behind. These moulds may hold enough detail to show external structure such as stomatal configuration and epidermal cell shape. Although sometimes difficult to interpret three-dimensionally, they are, with a trained eye, identifiable.

Petrification

Petrifaction is the result of replacement of some or all original organic material, cell for cell, by minerals dissolved in groundwater. If petrification is rapid enough, even soft tissues such as seed embryos can be preserved. Silica, calcite, or phosphates make up the bulk of minerals seen in petrified plants in the Drumheller valley.

Silicification, calcification, and phosphatization can occur in the same site or specimen and form in a variety of lithologies from mudstones to sandstones.

Silicification

Many plant fossils in the Drumheller area are partially or completely replaced by silica. The silica may only in-fill the open cell bodies with the cell walls still containing organic traces; or may completely replace them leaving no organic trace at all. Silicified specimens have a glassy texture and appear a clouded pale white in colour (Fig. 18). Silicified specimens usually show exquisite internal structure that can sometimes be seen with a hand lens.

Fig. 18. *Parataxodium* sp., silicified seed cone.

Calcification

Calcification is the most common form of plant replacement in the Horseshoe Canyon Formation. Calcite-replaced plants are soft and are commonly recessed from the surrounding rock due to weathering. Calcite reacts (bubbles) with citric or acetic acid (vinegar). Calcite-replaced plants usually have a pale dull yellow coloration (Fig. 19).

Phosphatization

Phosphatization does occur, although much more rarely than silicification or calcification. It can be seen as a dull gray sheen in sandstones in conjunction with carbonized remains. Phosphatization may appear massive and usually obliterates cell details due to the infilling, total replacement, and alteration during fossilization.

Fig. 19. *Parataxodium* sp., calcified seed cone containing seeds.

Carbonization

Carbonization is the result of pressure or heat from external physical forces, which drives out the organics in the plant and leaves behind carbon or charcoal. All of the surrounding coal seams in Drumheller are an example of compressed carbonized plant remains. Although fossilized, they are not petrifactions.

Carbonized plants are the most common form of plant remains found in the area. When found individually they are soft, friable, and black. They weather quickly on exposure to the elements. Many angiosperm leaf slabs contain the imprint of leaves with the carbon trace weathered away.

Organic trace is the term used to define more complex organic molecules than just mere carbon. Many of the preserved plants still contain complex organic molecules due to insufficient heat or pressure during preservation to totally carbonize the specimen. Both carbonized and organic trace remains can occur in the same specimen.

Fusinite is the common name given to preserved fusain or charcoal (burnt wood) pieces found throughout the sediments (Fig. 20). These fusinites are the remnants of Cretaceous forest fires that once burned sporadically across the floodplain. Fusain or charcoal (mere carbon) is the main component of these fossils. Fusinites do not compress during fossilization but can shatter and are sometimes preserved as such. Studies of fusain are rapidly catching on due to the excellent detail afforded by these fossils. In other formations charcoal fossils include flowers as well.

Fig. 20. Naturally weathered fusinites (black pieces).

Collecting Specimens

Before going out and collecting fossils, it can never be stressed enough to obtain landowner permission. One should never trespass onto private property. Collecting in national or provincial parks is forbidden unless you have a research permit. Each province in Canada has its own laws regarding the collecting of fossils on private and public lands. Alberta law states all fossils found in the province belong to the Crown. In Alberta, unless you have a permit, you are only allowed to surface collect. For more information, contact the Resource Conservation Manager at the Royal Tyrrell Museum at: 1-403-823-7707.

Upon discovering a fossil plant site, it is important to take a random sampling of specimens. This allows persons who view the collection an overall understanding of the site. Of great importance is the collection of any reproductive material, whether in the form of fern pinnules with sori, conifer cones, or angiosperm flowers or seedpods. These along with the foliage allow not only the ability to place the plant remains more easily into a family but may give evolutionary information as well.

Site locality information should be as detailed as possible. Date, collector, lithology (sediment source), terrain, elevation, landmarks, legal description, map co-ordinates, photographs, and geological formation, if available, should all be recorded. These allow others to locate your finds at a later date should more specimens be required. This information should be placed in your information catalogue.

Since fossil plants occur in a variety of lithologies, care must be taken in their collection to limit any damage that may occur during transport. Most consolidated specimens need little more than wrapping up with paper to prevent abrasion on other specimens. Specimens found in unconsolidated matrix such as shale may need to be preserved prior to wrapping and transport. Many acetone-based preservatives are available on the market such as Acryloid®, Butvar® or Vinac®. Acetone-based preservatives are preferred due to their fast drying ability. Acetone should be handled with caution as it is highly flammable liquid and vapour; it is harmful if inhaled and is an irritant to skin, eyes, and respiratory tract and it affects the central nervous system.

Any preservative used should be able to be removed later with the least amount of damage to the specimen so that it can be studied.

Cataloguing, Preparation, and Study of Specimens

After the specimens have been safely transported to the lab, care must be taken in their unwrapping. Specimens should be checked for damage and repaired. Depending on their preservation, some may need cleaning. This can be done by just dusting them off or actually washing them in water. At this time the specimens must be laid out and catalogued. The cataloguing of your specimens cannot be over-emphasized, whether you are an amateur or professional collector. You may have the greatest find of the century in your collections but without adequate information it is just another pretty rock. With this in mind, you should remember that not all finds are necessarily new to science, but they are still important as they may show structural variations or associations not seen before.

Methods of study are as varied as the types of preservation. Depending on the type of study, mechanical preparation with the use of dental probes or air-scribes may be employed. Chemical preparation is also quite common as well and may involve many types of acids or bases. Study not only may involve a broad overview of the specimen but may require the use of a light-microscope, Scanning Electron Microscope (SEM), or Transmission Electron Microscope (TEM) to view cellular details.

Many problems exist for the researcher trying to identify fossil plants. To this end, they make use of as wide a database as possible, which may involve many literature searches. An unknown specimen may present a variety of problems. A few are listed:

Problems Encountered with Interpretation of Fossil Specimens

Taxonomy, in this case the naming of fossil plants, is sometimes a problem due to limitations of specific, generic and familial traits. Defining these traits in living plant groups is sometimes difficult or confusing; the addition of fossil taxa sometimes compounds the problem. Fossils rarely represent the entire plant. Individual or multiple plant parts such as leaves, flowers, seeds, seed cones, pollen cones, or wood are used mainly to identify fossil plant genera and species. This limited information source can cause problems for the taxonomist. Other problems can be encountered as well.

Extinction

No living family, genus, or species is present for comparison. This is a common occurrence as one goes further back in geological time. Many Cretaceous forms, especially when dealing with angiosperms, have no living family or genus relatives.

Restriction of species morphological types

A living genus may have included various environmentally produced plant species in the past. Many species may have become extinct leaving only one living species. For example, *Sequoiadendron* survives with only one species, *Sequoiadendron giganteum*, which has scale-like leaves. In the past there may have been a myriad of species forms from linear leaved to scale-like or even deciduous representatives that have subsequently become extinct for unknown reasons, leaving only the single surviving species.

Expansion of genus morphological types

Just as restriction of species morphological types is a problem, so is the expansion of a plant type. Although the fossil genus or species may be extinct, its family may still be living but with many different morphological traits. The fossil may not be recognized as being part of a specific family or genus and may even be given a new family or genus name with no obvious connection. The same fossil is sometimes found to be related to its living family or genus at a later date.

Variability of preservation in fossils

As was discussed earlier, many forms of fossilization are encountered in the formation. This may cause problems of identification of identical forms due to different structures being either enhanced or degraded during fossilization and the subsequent re-description of the same structure as a new form.

Classification of Fossil Plants Found in the Horseshoe Canyon Formation

The following is a classification of plants represented by fossils from the Horseshoe Canyon Formation. This classification excludes plant groups that do not occur as palynomorphs or plant remains in the formation.

KINGDOM PROTISTA (ALGAE)
    Division Chlorophyta — green algae
    Division Chrysophyta — golden brown algae, diatoms
    Division Pyrrophyta — dinoflagellates
    Division Phaeophyta — brown algae
    Division Rhodophyta — red algae
KINGDOM MYCETEAE (FUNGI)
    Division Thallophyta
KINGDOM PLANTAE (PLANTS)
    Division Hepatophyta
        Family Sphaerocarpaceae
    Division Bryophyta — mosses
      Class Bryopsida
    Division Lycopodiophyta — club mosses and tassel ferns
      Class Lycopodiopsida
        Order Lycopodiales
          Family Lycopodiaceae — club mosses
            Genus *Phylloglossum*
        Order Selaginellales — club mosses and spike mosses
        Order Isoetales
          Family Isoetaceae — quillworts
    Division Sphenophyta
      Class Equisetopsida
        Order Equisetales
          Family Equisetaceae — horsetails
            Genus *Equisetum*
    Division Filicophyta — true ferns
      Class Filicopsida
        Order Salviniales
          Family Azollaceae — mosquito fern

Genus *Azollopsis*

Family Salviniaceae — water spangles

Genus *Dorfiella*

Order Marsileales

Family Marsileaceae — water clovers

Order Osmundales

Family Osmundaceae — Osmunda

Genus *Osmunda*

Order Schizaeales

Family Schizaeaceae — curly grass

Order Polypodiales

Family Blechnaceae

Family Dennstaedtiaceae — cup fern

Order Gleicheniales

Family Gleicheniaceae

Family Matoniaceae

Order Cyatheales — tree fern

Order Ophioglossales

Family Ophioglossaceae — succulent fern

Division Cycadophyta

Order Caytoniales

Family Caytoniaceae — seed ferns

Order Bennettitales — false cycads

Division Ginkgophyta

Order Ginkgoales

Family Ginkgoaceae — maidenhair tree

Genus *Ginkgo*

Division Coniferophyta

Class Coniferopsida

Order Coniferae

Family Pinaceae — pines

Family Taxodiaceae — redwoods

Subfamily — Taxodioideae

Tribe — Sequoieae

Genus *Parataxodium* — deciduous *Sequoiadendron*

Tribe — Cryptomerieae

Genus *Taxodium* — swamp cypress

Tribe — Cunninghamieae

Genus *Cunninghamia* — china fir
Genus *Taiwania*
Tribe — Sequoieae
Genus *Athrotaxis*
Family Cupressaceae — cypresses
Genus *Mesocyparis*
Family Podocarpaceae — yellow woods
Class Taxopsida
Order Taxales
Family Taxaceae — yews
Genus *Torreya*
Division Angiospermophyta
Class Magnoliopsida (Dicotyledonae)
Order Laurales
Family Lauraceae — laurel
Order Nymphaeales
Family Nymphaeaceae — lotus
Order Ceratophyllales
Family Ceratophyllaceae — hornworts
Genus *Ceratophyllum*
Class Eudicotyledonae
Order Saxifragales
Family Cercidiphyllaceae — katsura
Family Haloragaceae
Family Hamamelidaceae — witch hazel
Order Proteales
Family Platanaceae — plane tree
Order Trochodendrales
Family Trochodendraceae — wheel tree
Order Rosales
Family Ulmaceae — elm
Order Fagales
Family Myricaceae — myrtles, bayberry
Family Fagaceae — beech
Family Betulaceae — birch
Family Carpinaceae — hornbeams
Order Malpighiales
Family Salicaceae — willow

Order Ericales
    Family Symplocaceae — sweetleaf
Order Caryophyllales
    Family Amaranthaceae
Order Gunnerales
    Family Gunneraceae
Order Myrtales
    Family Lythraceae
    Family Nyssaceae
        Genus *Nyssa*
        Genus *Davidia*
Order Santalales
    Family Loranthaceae — mistletoe
Order Aquifoliales
    Family Aquifoliaceae — holly
Order Buxales
    Family Buxaceae — boxwood
Order Sapindales
    Family Anacardiaceae — sumac, cashew
    Family Simaroubaceae — tree or heaven
Order Apiales (Umbellales)
    Family Toricelliaceae
    Family Apiaceae (Umbelliferae) — celery or parsnip
Order Lamiales
    Family Oleaceae — olive
Class Monocotyledonae (Liliopsida)
    Order Alismatales
        Family Araceae — arums
            Genus *Albertarum*
    Order Arecales
        Family Arecaceae — palms
    Order Liliales
        Family Liliaceae — lilies
    Order Poales
        Family Cyperaceae — sedges

THE HISTORY OF FOSSIL PLANT COLLECTING WITHIN THE HORSESHOE CANYON FORMATION OF DRUMHELLER, ALBERTA

The written history of fossil plant collecting in the Horseshoe Canyon Formation is long but sparse. In 1908, Penhallow described the first fossil plants from the Lower Edmonton Formation (Horseshoe Canyon Fm.). These fossils were named:

Cupressinoxylon macrocarpoides (Penhallow 1904)
Sequoia albertensis
Picea albertensis.

Both *C. macrocarpoides* and *S. albertensis* are wood samples with *P. albertensis* representing a seed cone. Both wood samples were placed in the Taxodiaceae with *C. macrocarpoides* identified as *Cupressus*-like and *S. albertensis* as *Sequoia*-like. *P. albertensis* was identified as being a *Picea* cone but was not figured.

In 1937 Shimakura renamed *Sequoia albertensis* as *Taxodioxylon albertensis*.

In 1949 the best and most popular account of fossil plants from the Upper Cretaceous and Paleocene of Alberta was written by W. A. Bell while working for the Geological Survey of Canada. The monograph "Uppermost Cretaceous and Paleocene floras of western Canada," Bulletin 13, is still commonly quoted today. Bell listed eighteen species of plants from the lower Edmonton Formation (Horseshoe Canyon Fm.). The following is a listing of these plants placed in the appropriate family or genus based on investigations by the present author.

FOSSIL GEN. ET SP.	PLANT PART	FAMILY OR GENUS REPRESENTED
Pityostrobus (Cunninghamiostrobus?) sp.	Seed cones	Taxodiaceae, *Cunninghamia*
Torreyites tyrrellii (Dawson) Bell	Leaves and branches	Taxodiaceae, mixed unknowns and *Cunninghamia*
Dombeyopsis nebrascensis (Newberry) Bell	Leaves	Rhamnaceae?
Elatocladus intermedius Hollick	Foliage	Taxodiaceae, *Parataxodium*
Juniperites gracilis (Heer) Seward and Conway	Pollen cones	Taxodiaceae, *Parataxodium*
Sequoiites artus Bell	Foliage	Taxodiaceae, *Parataxodium*
Sequoiites dakotensis Brown	Seed cones	Taxodiaceae, *Parataxodium*
Equisetum perlaevigatum Cockerell	Stems	Equisetaceae, *Equisetum*
Ginkgoites sp.	Leaves	Ginkgoaceae, *Ginkgo*
Jenkinsella arctica (Heer) Bell	Seed pods	Cercidiphyllaceae
Trochodendroides arctica Heer	Leave	Cercidiphyllaceae
Nilsonia serotina Heer	Leaves	Nilsoniaceae?
Nilsonia sp.	Leaves	Nilsoniaceae?
Nymphaeites angulatus (Newberry) Bell	Leaves	Trapaceae
Thuites interruptus Newberry	Foliage	Cupressaceae, *Mesocyparis*
Vitis stantoni (Knowlton) Brown	Leaves	Platanaceae
Carpolithus (Ginkgoites?) fultoni Bell	Seeds	Family indet.
Carpolithus (Ginkgoites?) kneehillensis Bell	Seeds	Family indet.

After 1949 a large gap occurred without investigations into the flora until 1965 when W. A. Bell published "Illustrations of Canadian Fossils: Upper Cretaceous and Paleocene Plants of Western Canada," Paper 65–35. This included three plants from the area of which all, *Vitis stantoni*, *Dombeyopsis nebrascensis*, and *Sequoiites dakotensis*, were already known from the formation. This paper was more of a pictorial account without discussion or descriptions.

Ramanujam and Stewart (1969a) named four fossil woods identified as belonging to the Taxodiaceae. These wood types are listed below with their closest affinity as per the original author's identification.

FOSSIL GEN. ET SP.	FAMILY OR GENUS REPRESENTED
Taxodioxylon antiquium	Unknown
Taxodioxylon drumhellerense	Taxodiaceae, *Glyptostrobus*
Taxodioxylon gypsacum (Goppert) Krausel 1949	Taxodiaceae, *Sequoia*
Taxodioxylon taxodii Gothan 1905	Taxodiaceae, *Taxodium*

Slowly research into plants of the area increased. In 1986 Muhammad identified two more Taxodiaceous woods from the area. These were:

FOSSIL GEN. ET SP.	FAMILY OR GENUS REPRESENTED
Taxodioxylon sp. (Hortig) Gothan 1905	Taxodiaceae, *Taiwania?*
Taxodioxylon gypsacum (Goppert) Krausel 1949	Taxodiaceae, *Sequoia*

Both identifications were based on Ramanujam and Stewart (1969a). Muhammad also noted the similarity of *Taxodioxylon* sp. (Hortig) Gothan 1905 to that of *Taxodioxylon multiseriatum* from near Medicine Hat, Alberta.

In 1991 Serbet and Stockey discovered Taxodiaceous pollen cones that were named *Drumhellera kurmanniae*. The specimens were preserved as calcite and silica and entombed in ironstones as well as carbon trace preserved in sandstone. These plants were suspected as belonging to the Sequoias. These are identical to *Juniperites gracilis* Bell (1949).

At the same time, in 1991 Aulenback and Braman developed a new acid etching technique for investigating fossils preserved in ironstones, which sparked further interest in plants from the formation.

In 1994 McIver and Aulenback described a cupressid, *Mesocyparis umbonata* from specimens recovered from ironstone.

In 1997 Serbet completed his PhD thesis on the fossil plants of the Horseshoe Canyon. The unpublished thesis was titled "Morphologically and anatomically preserved fossil plants from Alberta, Canada: A flora that supported the dinosaur fauna during the Upper Cretaceous (Maastrichtian)."

Although a larger flora than Bell's was identified, some can be seen as re-named genera and species, previously described by Bell. The following is a listing of plants from Serbet's thesis excluding the miscellaneous woods and roots. Based on recent investigations their closest affinities are listed. A minimum of twenty-three genera are represented:

FOSSIL GEN. ET SP.	PLANT PART	FAMILY OR GENUS REPRESENTED
Jungermanniales	Foliage	Azollaceae
Moss gametophyte	Foliage	Azollaceae
Lycopodiaceae	Megaspore	Lycopodiaceae
Osmunda cinnamomea	Rhizomes	Osmundaceae, *O. cinnamomea*?
Microlepia sp. Type A, *Microlepiopsis bramanii*	Rhizomes	Dennstaedtiaceae? (Schizaeaceae?)
Microlepia sp. Type B, *Microlepiopsis aulenbackii*	Rhizomes	Dennstaedtiaceae? (Schizaeaceae?)
Blechnum sp. A, Chap. 2, *Blechnum nishidae*	Rhizomes	Blechnaceae? *Woodwardia*?
Blechnum sp. B, Chap. 2, *Blechnum oroszii*	Rhizomes	Blechnaceae?
Nilsonia sp.	Leaves, seed cones	Nilsoniaceae?
Ginkgo sp.	Leaves, pollen cone	Ginkgoaceae, *Ginkgo*
Cunninghamia sp.	Seed cones, foliage	Taxodiaceae, *Cunninghamia*
Drumhellera kurmanniae Serbet and Stockey 1991	Seed cones, pollen cones, foliage	Taxodiaceae, *Parataxodium* sp.
Taxodium sp.	Seed cones, pollen cones, foliage	Taxodiaceae, *Taxodium*
Mesocyparis umbonata McIver and Aulenback 1994	Seed cones, pollen cones, foliage	Cupressaceae sp.
Taxus sp.	Leaves	Taxaceae, *Torreya*

Picea sp.	Seed cone bracts	Pinaceae, *Picea*
Cercidiphyllaceae	Leaves, seed pods	Cercidiphyllaceae
Trochodendraceae #A	Fruiting axis	Trochodendraceae
Trochodendraceae #B and seed type G	Fruiting axis	Trochodendraceae
Betulaceae, *Carpinus* sp.	Nutlets	Betulaceae, *Carpinus*
Cyperaceae	Seeds	Cyperaceae
Porosia verrucosa = *Carpites verrucosus* (Lesquereux) McIver and Basinger 1993	Seeds	Unknown
Seed type A	Fruits	*Ilex*-like
Seed type B	Seeds	Unknown
Seed type C, N, and O	Drupes	Davidiaceae, *Davidia*
Seed type D, E, I, J, and L	Seeds	Nyssaceae
Seed type F	Seeds	Araliaceae, *Toricellia*
Seed type H	Seeds	Unknown
Seed type K	Fruiting structure	Unknown
Seed type M	Seed pods	Unknown
Seed type P	Plant part	Unknown
Seed type Q	Tubers	Equisetaceae, *Equisetum*
Seed type R	Seeds	Unknown

In 1998 *Taxodium wallisii* was described by Aulenback and LePage. This was the first accurately identified *Taxodium* in the formation.

In 1999 an *Osmunda cinnamomea* was described by Serbet and Rothwell based on Serbet's 1997 thesis.

In 2003 *Microlepiopsis bramanii* and *M. aulenbackii* were also published based on the thesis material of Serbet (1997) by Serbet and Rothwell (Serbet 1997; *Microlepia* sp. Type A = *Microlepiopsis bramanii* Chap. 2, *Microlepia* sp. Type B = *Microlepiopsis aulenbackii* Chap. 2).

In 2005 Bogner, Hoffman, and Aulenback described *Albertarum pueri*, an ancient Araceae related to living *Symplocarpus*, the skunk cabbage.

In 2006 *Midlandia nishidae* and *Wessiea oroszii* were formally described by Serbet and Rothwell based on the specimens from Serbet's thesis (Serbet 1997; *Blechnum* sp. B = *Blechnum oroszii* Chap. 2, and *Blechnum* sp. A = *Blechnum nishidae* Chap. 2).

In the formation a Cycadophyte is presently being described by Aulenback from seed cones belonging to an extinct family of the Bennettitales. As well, the author is presently describing four fossil ferns, one from the Marsileales and three from the Salviniales.

A large amount of collected unidentified material from different lithologies exists from the Horseshoe Canyon. As will be seen, these present listings are by no means the end of this diverse floral assemblage, and many that are shown are presently unnamed. Ongoing investigations into the flora have produced many new discoveries and insights.

PRESENT STATUS OF THE PLANT FOSSIL RECORD

PREVIOUSLY DESCRIBED AND UNDESCRIBED FOSSIL PLANTS AND THEIR AFFINITIES

Many plants do not leave behind any fossil plant trace either due to decay or poor potential for fossilization (living in upland areas, etc.). They are identified only due to the presence of pollen (gymnosperms and angiosperms) or spore (cryptogams: ferns, mosses, fungi, etc.). Pollen and spores are very resistant to decay during fossilization except in well-oxygenated environments.

In the fossil record pollen and spores are treated as form taxa and are named as such. This name usually has no bearing on its relationship to the actual plant producing the palynomorph, although many pollen and spore types are referable to recent families. The spores and pollen identified in the Horseshoe Canyon Formation have been placed into appropriate families and genera where possible. Many spore and pollens are not identifiable to recent families and are listed as unknowns. Even when placed into families, many palynomorphs cannot be related to genera or species.

The palynomorphs included in this plant listing are a compilation based on the works of Srivastava (1970) and Jarzen (1982) and synthesized by Dr. D. Braman of the Royal Tyrrell Museum with many additions made by him over the years.

With the merging of the identified palynomorphs and the named and un-named plant discoveries, a listing has been developed which, although even more comprehensive, is by no means complete. Where present, the palynomorphs have been included to show or fill gaps in the plant record. Some plant genera that were not represented by palynomorphs are included due to their presence as fossils. Although they do not have a recognized palynomorph representative, it may be as yet one of the unknowns that are listed.

Discrepancies are discussed in both the palynomorphs and plants. The majority of the plants are as yet scientifically undescribed.

The taxonomy for the cryptogams and ferns is developed from Jones (1987) and Smith et al. (2006). The Salviniales/Marsileales is retained and explained based on the fossil record.

The taxonomy used for the gymnosperms is adapted from Krussmann (1991). Although most authors now merge the Taxodiaceae/Cupressaceae, they are dealt with separately.

The taxonomy for the angiosperms is adapted from Krussmann (1986) and Stevens (2001).

All fossils shown are from the Horseshoe Canyon Formation, Drumheller area unless specified in their figured caption or text.

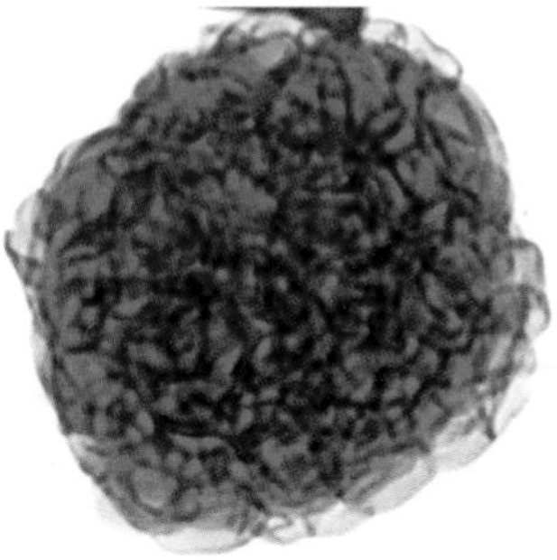

Fig. 21. *Palambages canadiana.*

PROTISTA (ALGAE)

Chlorophyta, green algae

Chrysophyta, golden brown algae, diatoms

Pyrrophyta, dinoflagellates

Phaeophyta, brown algae

Rhodophyta, red algae

At present the relationship of the spore types (Fig. 21) to any living or extinct family is unknown and identifiable plant remains have not been found. Three extant algae are figured (Figs. 22–24).

Fossil spore:
Palambages canadiana
Schizosporis cooksoni
Schizosporis parvus
Schizosporis sp.

Fig. 22. An extant Rhodophyta.

Fig. 23. An extant Phaeophyta.

Fig. 24. An extant Chlorophyta.

MYCETEAE (FUNGI)

Unusual fossil remains have been found in the form of masses of small spiculate sphericals. These appear to the naked eye as minute yellowish to whitish masses embedded in ironstones (Fig. 25). When etched in acids, these sphericals are found either encircling small branches (Fig. 27) or on leaf surface fragments forming mats (Fig. 26). Although they appear egg-like in form, they are not.

The spheres are minute, less than 1 mm in diameter and, in better preserved specimens, contain a netted dark mass internally (Figs. 28–30). None of the spheres have been found to contain spores. The sphere-netted, dark mass sometimes contains a small elongate papillae (Fig. 29). The external spicules that cover them are less than 0.25 mm long, hollow, slender, pointed, and opened at both ends (Fig. 32). The spicules apparently stood erect on the surface of the sphere based on the broken bases found on many spheres.

The sphericals are tentatively identified as fungi as one specimen shows a filament-like fungal hyphae connecting two spheres (Fig. 30).

There may eventually prove to be more than one organism present as small silicified bean-shaped cells external to the spheres and mixed towards the bases of the spicules are quite common (Fig. 31). This fossil remains unstudied and un-named.

Fig. 25. Weathered sphericals in ironstone.

Fig. 26. Ground thin-section showing internal dark mass in silica sphericals.

Fig. 27. Sphericals encircling a small branch.

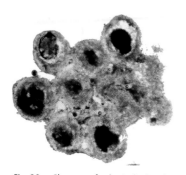

Fig. 28. Close-up of sphericals showing structure.

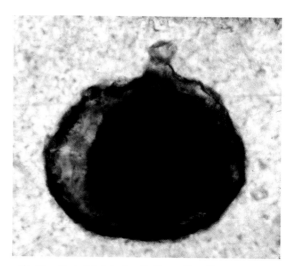

Fig. 29. Central dark body with the papillate-like extension.

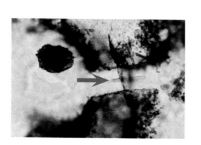

Fig. 30. Fungal hyphae bridge (arrow).

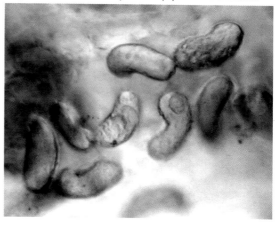

Fig. 31. Bean-shaped silica remains from the surface of a spherical.

Fig. 32. Hollow spicule with air trapped within the centre. The basal attachment to the spherical is to the left.

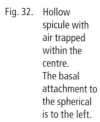

Thallophyta

Fungi have both a vegetative and a reproductive phase. The reproductive phase is usually the more visual portion, as is seen in the mushroom or other various forms (Figs. 33–35). The vegetative phase is usually unseen in the form of fungal hyphae.

Vegetative remains of fungi in the formation are found in preparations of thin sections of silicified plant material. The fungi are in the form of microscopic hyphae in the leaves (Fig. 37) and wood (Fig. 36) and most likely represent post-depositional decay of the vegetation prior to fossilization. Dispersed fragments of fungal remains are found in palynological preparations (Fig. 38).

It is likely that many different fungi occurred in the environment and that this sampling of spore and vegetative material represents only a minor part of the diversity present.

Fossil spore:
Pluricellaesporites clarkei
Pluricellaesporites elongatus
Pluricellaesporites glomeratus
Pluricellaesporites nodosus

Fig. 33.   *Scleroderma* sp., earthball fungus.

Fig. 34.   *Ramaria botrytis*, golden coral fungus.

Fig. 35.   *Clitocybe* sp., funnelcap mushrooms.

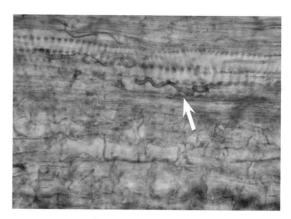

Fig. 36. Unidentified fossil angiosperm wood with fungal hyphae within the cells.

Fig. 38. *Pluricellaesporites* sp.

Fig. 37. *Mesocyparis umbonata* leaves with fungal hyphae in tissue (arrow).

PLANTAE

Hepatophyta

The Hepatophyta or liverworts are considered lower plants and inhabit moist environments worldwide.

Although a large amount of fragments of thalli are found in certain palynological preparations, no identifiable mega-plants have yet been recovered. There are a few spore species known (Fig. 39), which may give hope for future discoveries of some interesting plants.

Fossil spore:
Hannisporis scollardensis
Triporoletes asper
Triporoletes stellatus
Triporoletes tornatilis
Triporoletes tympanoideus
Triporoletes sp.
Trochicola scollardiana

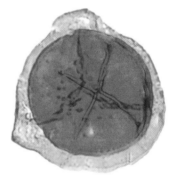

Fig. 39.   *Triporoletes stellatus* spore.

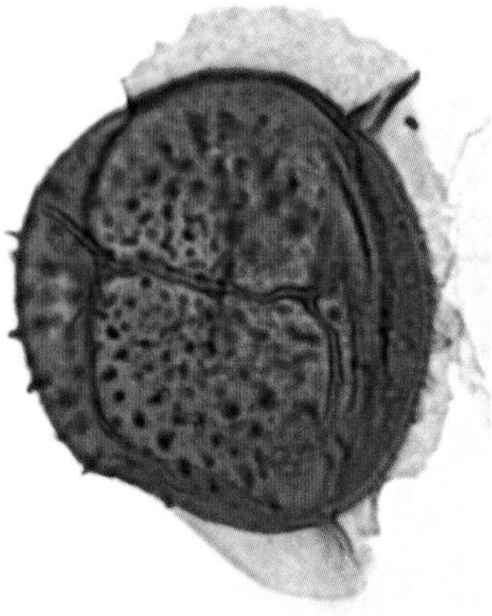

Fig. 40. *Aequitriradites spinulosus* spore.

SPHAEROCARPACEAE

Very few spore species are found representing the hornworts and liverworts (Fig. 40), and plant remains are unknown in the formation. Liverworts and hornworts require the same growing conditions as ferns. The living genus *Marchantia* is most commonly encountered by gardeners (Figs. 41, 42). In the horticultural trade, liverworts, such as *Marchantia*, are sometimes considered invasive. Liverworts and hornworts enjoy worldwide distribution.

Fossil spore:
Aequitriradites ornatus
Aequitriradites spinulosus
Aequitriradites verrucosus

Fig. 41. *Marchantia* sp., plant with antheridiophores and gemmae cups.

Fig. 42. *Marchantia* sp., archegoniophores.

Bryophyta

BRYOPSIDA (MOSSES)

Mosses (Fig. 43) are rare in any fossil flora and the Horseshoe Canyon Formation is no exception. Cretaceous fossil plant remains of mosses have not been identified in the field, but two vegetative stem fragments assignable to the Bryophyta have been collected as silicified remains from processed ironstones. These fossils are small, up to 10 mm long, and retain both external and internal anatomy (Figs. 44–46). Even with this information, it would be difficult to place these fossils in a living genus. Very few spore taxa are found (Fig. 47).

Many genera occur worldwide with all preferring cool, wet moist semi-shade to shade conditions. Many living species can be found in Alberta. It can be seen that even spore genera are not numerous, but studies are ongoing and the species list will definitely increase in the future.

Fossil spore:
Cingutriletes clavus
Stereisporites antiquasporites

Fig. 43.   *Polytrichum commune.*

Fig. 44.   Fossil moss.

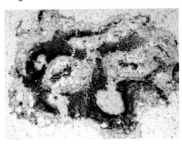

Fig. 45.   Cross-section of a bifurcate axis showing internal cell structure in the fossil moss.

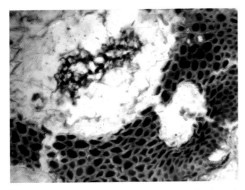

Fig. 46.   A close-up from Fig. 45 of a central stele and departing leaf trace.

Fig. 47.   *Stereisporites antiquasporites* spore.

Fig. 48. *Lycopodium annotinum,*
running pine.

Fig. 49. *Lycopodium complanatum,*
ground cedar.

Lycopodiophyta

LYCOPODIOPSIDA

Lycopodiales
(Club mosses and tassel ferns)

The Lycopodiaceae contains up to a thousand recent species in four genera: *Cernuistachys, Lycopodium, Selaginella,* and *Isoetes.* Most are creeping plants clothed in small scale-like leaves. Plants are heterosporous with reproductive parts sometimes held erect forming cones (Figs. 48, 49).

The club mosses contain rich spore diversity in the formation (Fig. 50), but identifiable plant remains have not been found.

The Lycopodiaceae are cosmopolitan in distribution with many tropical, temperate, and subarctic forms. All species rely on commensalistic mycorrhizal fungi for growth and prefer acidic soil conditions.

Fossil spore:
Camarozonosporites concinnus
Camarozonosporites insignis
Echinisporites sp.
Hamulatisporis albertensis
Hamulatisporis loeblichii
Hamulatisporis rugulatus
Lycopodiacidites triangularis

Lygodioisporites verrucosus
Retitriletes austroclavatidites
Retitriletes coulianus
Retitriletes lucifer
Retitriletes mirabilis
Retitriletes muricatus
Retitriletes nidus
Retitriletes papillaesporites
Retitriletes reticulumsporites
Retitriletes reticulisporites
Retitriletes saxatilis
Retitriletes singhii
Trilites bettianus

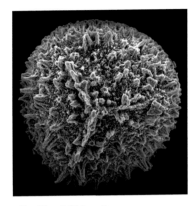

Fig. 50.   *Echinisporites* sp., spore.

Phylloglossum

These are some of the few spores (Fig. 51) identified to a living genus, although fossil remains are
unknown in the formation. Living *Phylloglossum*
is a tiny terrestrial, monosporous Lycopodiaceae
with a basal group of leaves arranged in a spiral
with a short stalk ending in a terminal strobilus.
The genus is monotypic (*Phylloglossum drummondii*) and is restricted to eastern Australia.

Fossil spore:
Foveolatisporites sp.
Microfoveolatosporis skottsbergii
Reticulosporis foveolatus

Fig. 51.   *Reticulosporis foveolatus* spore.

Fig. 52. *Selaginella uncinata*, in cone (arrow).

Fig. 53. *Selaginella kraussiana*, vegetative phase. Sporangia are produced in the axils of the leaves.

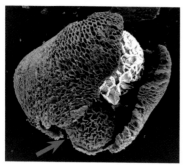

Fig. 54. *Selaginella*-like megasporangia bearing *Erlansonisporites spinosus* mature and abortive (arrow) megaspores.

Selaginellales (Club mosses and spike mosses)

The fossil record of this order presently only consists of spores (Fig. 55) in the formation and a single recovered megasporangium (Fig. 54). In living species, sporangia can be produced in defined cones (Fig. 52) or in the axils of leaves (Fig. 53).

The order contains only one family and one genus *Selaginella* with up to seven hundred species. Most *Selaginella* prefer moist to wet acidic soils in shady conditions.

Fossil spore:
Acanthotriletes varispinosus
Ceratosporites couliensis
Ceratosporites equalis
Ceratosporites morrinicolus
Ceratosporites pocockii
Ceratosporites sp.
Erlansonisporites spinosus
Heliosporites altmarkensis
Heliosporites kemensis
Lusatisporis dettmannae
Neoraistrickia speciosa
Neoraistrickia truncata
Pilosisporites sp.
Raistrickia artebla

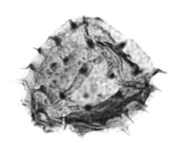

Fig. 55. *Heliosporites kemensis* spore.

Isoetales

ISOETACEAE (QUILLWORTS)

This family consists of 150 or more recent species divided into two genera, *Isoetes* and *Stylites*. These heterosporous plants consist of linear grass-like leaves held in a helical rosette upon a corm (Fig. 56). The basal swollen portion of the leaf contains the reproductive organs, which consist of both mega and microsporangia.

Fig. 56. *Isoetes* sp., quilwort. Reproduced from Brown (1935).

Corms of *Isoetes* are highly diagnostic and are found as fossils in other Cretaceous and Paleocene formations (McIver and Basinger 1993), but, unfortunately, none to date have been recovered from the Horseshoe Canyon Formation. Presently, only spores represent this order in the formation (Fig. 57). It is hoped that they will eventually be found as the spores appear common, are diverse in form, and may represent multiple species in the flora.

Living quillworts prefer slightly acidic waters on the margins of lakes or low swampy areas in full sun. The *Isoetes* range from cool temperate to tropical.

Fossil spore:
Minersisporites delicatus
Minersisporites cf. *deltoides*
Minersisporites mirabilis
Ricinospora cryptoreticulata
Minerisporites deltoids X M. mirabilis

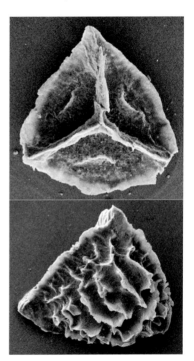

Fig. 57. *Minerisporites deltoids X M. mirabilis* megaspore, proximal (top) and distal (bottom) polar views.

Fig. 58.   *Equisetum arvense,*
reproductive structure.

Sphenophyta

EQUISETOPSIDA

Equisetales

EQUISETACEAE (HORSETAILS)

Equisetum is the only living genus of the Sphenophyta and has approximately fifteen species that are divided into two subgenera, *Equisetum* and *Hippochaete*. They are homosporous plants that consist of linear rhizomes, jointed stems, whorled leaves, and branches with a terminal cone on fertile shoots.

Although *Equisetum* has been recorded from the Carboniferous, these identifications are thought to be doubtful with true *Equisetum*-like plants speculated to have evolved by at least the Triassic (Taylor 1981). Intact rhizomes in growth position are known from the Jurassic age, Morrison Formation of Utah. In Alberta, isolated *Equisetum*-like storage organs and stems are found in the Oldman and Dinosaur Park Formations.

As a very common fossil at many sites, *Equisetum* can be found as isolated aerial nodes (Fig. 61) and stem fragments as well as rhizome sections with or without tubers (Figs. 62–64). Fragments can be silicified, calcified, or mud-cast in-fill material.

At least two distinct forms can be recognized in the flora; one with exceptionally large rhizomes and tubers, and another type with exceedingly small rhizomes and tubers.

The tubers of the smaller rhizomes are small and spherical, up to 10 mm in diameter and appear to have been fleshy (Fig. 62). The tubers are reminiscent of fruits described as Nymphaeaceae sp. from the Cretaceous of Germany (Knobloch and Mai

1983), *Myristica catahoulensis* from the Paleocene of southwestern Saskatchewan (McIver and Basinger 1993), or *Striatisperma coronapunctatum* from the Eocene of Oregon (Manchester 1994). These fossils seem co-specific.

The large rhizomes contain tubers up to 2.3 cm long and are elliptical to oval in form (Fig. 64). Both large and small rhizomes can produce multiple tubers at the node. As fossils split along the bedding plain, usually only one or two tubers are exposed.

Certain living species of the subgenus *Hippochaete* produce underground tubers. *E. sylvaticum*, Wood Horsetail (Fig. 59) and *E. arvense*, Field Horsetail (Figs. 58, 60), are two North American species that regularly produce underground tubers. *E. arvense* produces a rounded tuber and *E. sylvaticum* produces elliptical tubers. It is possible that the fossil rhizome represents species similar to these two living forms.

Although common in the formation, neither fossil has been studied in depth and is rarely mentioned in fossil flora listings of the area. The larger fossil species was previously named *Equisetum perlaevigatum* (Bell 1949), while the smaller fossil is un-named. Other species may also be represented in the flora. The palynomorphs appear identical to those of *Gnetaceaepollenites* (compare Figs. 65 and 498), which may also belong to the Equisetaceae. Presently, *Equisetum* enjoys a worldwide distribution colonizing moist areas in full sun or shade.

Fossil spore:
Equisetosporites amabilis
Equisetosporites lajwantis
Equisetosporites menakae
Equisetosporites mollis

Fig. 59. *Equisetum sylvaticum*, bearing reproductive structures.

Fig. 60. *Equisetum arvense*, bearing the later seasonal growth of sterile foliage.

Fig. 61. Fossil *Equisetum* jointed section of branch.

Fig. 62. Fossil *Equisetum* rhizome section with tubers present.

Fig. 63. Fossil *Equisetum* large jointed rhizome section without tubers present.

Fig. 64. Fossil *Equisetum* large jointed rhizome section with tubers present.

Fig. 65. *Equisetosporites lajwantis* spore.

FOSSIL PLANTS

Filicophyta

FILICOPSIDA (FERNS)

The Filicophyta or true ferns can be found in the fossil record as far back as the Carboniferous (Tidwell 1998; Taylor 1981). All ferns reproduce by spores. The basic structural form of a fern is given in Figure 68.

The stems of ferns which grow in, or on, the ground are called rhizomes. The rhizomes bear the roots.

The frond consists of the stipe (stem) and blade (leaf). The blade can be broken down into the rachis and pinnae. The blade can be either simple or compound. A compound blade is divided into distinct pinnae (leaves) and may be termed either pinnately or palmately compound (Fig. 66). The rachis is the main axis from which the pinnae or lateral segments of the blade arise (Fig. 68).

Dissection of the blade into lateral segments results in pinnate, bipinnate, or tripinnate fronds (Fig. 67). Fronds can be vegetative or reproductive.

The pinnae can be divided into finer individual leaflets called *pinnules*. Commonly reproductive fronds have dots on the pinnule's dorsal side called *sori* (Fig. 68). A sorus (singular of sori) commonly consists of an indusium (protective covering) and groupings of sporangia. Ferns sometimes produce sori without indusium.

Sporangia are where spores are produced. Most sporangia contain a thickened area or row of cells called an annulus, which shrinks upon drying to release the spore. Spores are termed either *monolete* or *trilete* depending on their markings

Fig. 66.   Drawing examples of: A = pinnate and B = palmate fern fronds.

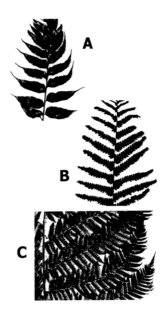

Fig. 67.   Drawing examples of: A = pinnate, B = bipinnate, and C = tripinnate dissection of fern fronds.

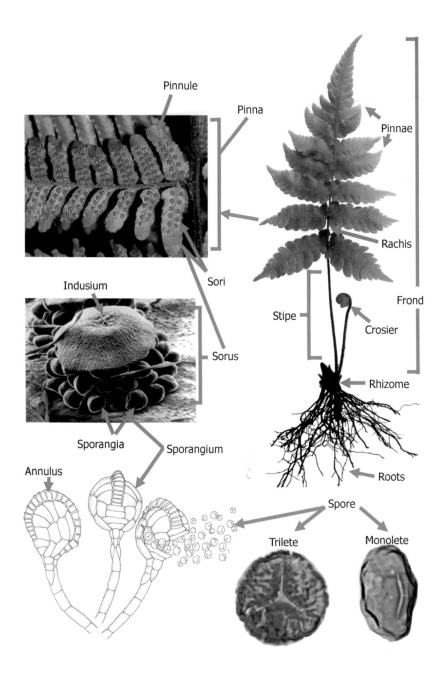

Fig. 68.   Simplified anatomical structures of a generalized fern.

(Fig. 68). Each fern species contains a specific spore type.

The majority of extant fern spores are called *miospores* and are, on average, from 20 to 40 μm in diameter. Some ferns are heterosporous and produce both mega- and microspores. Microspores can be from 10 to 60 μm in diameter and are termed micro- in relationship to the megaspore of the fern. Cretaceous and extant megaspores commonly range from 80 to 1,000 μm in size.

There are many variations in the shape and form externally and internally of the pinna, sori, indusium, sporangia, rhizomes, roots, and spores found in ferns. All of these morphological differences help define family, genera, and species.

Fig. 69.   *Azolla caroliniana* surface view.

Fig. 70.   *Azolla caroliniana* lateral view.

Salviniales

AZOLLACEAE (FILMY FERNS)

The Azollaceae are a family of diminutive, free-floating aquatic, heterosporous ferns (Figs. 69, 70). The living Azollaceae consists of the single genus *Azolla*, which is divided into two sections, *Azolla* and *Rhizosperma*. These sections are based on the number of massae (3 or 9) and the shape of the perinal prolongations (fluke or anchor shaped vs. pointed or blunt-tipped). The four species in section *Azolla* have three massae per megaspore and two sori per plant. In the section *Rhizosperma* are two species *A. nilotica* and *A. pinnata*. These species both contain nine massae per megaspore, with *A. nilotica* having four sori per plant instead of two. *A. pinnata* contains only two sori similar to section *Azolla*. Individual species are separated based on morphology of both megaplant and reproductive structures.

The Azollaceae is represented in the fossil record by as many as thirty Cretaceous species worldwide, based on megaspores and microspores and many more in the Cenozoic (Kovach and Batten 1989). *Azolla*-like megaspores originate around the Coniacian/Santonian boundary (85.8 mybp) with the divergence of both *Azollopsis* and the *Azolla conspicua* megaspore forms from *Ariadnaesporites*. Other closely related *Azolla*-like forms diverged from the *Azolla conspicua* form around the Santonian/Campanian boundary (83.5 mybp). It is speculated that true *Azolla*, based on both plant and megaspore form, do not appear until the early Tertiary (64 mybp).

Of the five fossil plant types possible, based on megaspores described from the formation, two *Azolla* megaspore *formae* have been found asso-

46

ciated with one foliage type and another *Azolla* megaspore *formae* with grouped mega- and microspores in a reproductive structure.

Megaspores and vegetative material of the whole plant fossil are commonly encountered in preparation of ironstones for silicified remains (Figs. 72, 73). This minute (< 5 mm) fossil plant is interpreted as containing a thin horizontal stipe with small crowded tri-lobed pinna and a long slender modified pinna to function for absorption, which will be referred to as a *pinna/root* (Figs. 71, 74).

The fossil stipe has a zigzag outline changing direction at each pinna. The tri-lobed pinna is differentiated into a reduced dorsal (upper) lobe; a larger, ovate to round, ventral (partially submerged) lobe, and a reduced median lobe (Figs. 71, 74–76).

Modified pinna/root sections up to 15 mm long and 1 mm wide are common in fossil preparation. Although many pinna/roots are found in association as well as attached, none represent a complete mature length. Some pinna/root preparations contain well-preserved internal anatomy (Figs. 79, 80).

Sori were produced in pairs (Figs. 81–84) as in modern *Azolla* (exclusive of *A. nilotica*). Intact attached megaspore sori with partially preserved indusium as well as isolated megaspores (Figs. 82–90) and microsporangia (Figs. 91, 92) have been recovered at various stages of development.

Sori contain individual mature megaspores from 300 to 400 μm long. Megaspores were previously given the form genus names *Azolla fistulosa* and *A. velus*, which contain from twenty to thirty massae. These megaspores show excellent external and internal structures (Figs. 85–90).

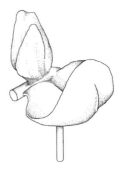

Fig. 71.   Drawing of a single plant of fossil Azollaceae.

Fig. 72.   Fossil Azollaceae in ironstone. Scale bar = 2 cm.

Fig. 73.   fossil Azollaceae acid etched sample.

Fig. 74. Fossil plant section bearing eroded lobes and pinna/roots.

Fig. 75. Fossil growing tip showing
          pinnae lobes.

Fig. 76. Fossil plant section showing
          internal vasculature.

Microsporangia are oval to round with up to eight massulae (Fig. 91). Perinal prolongations are restricted to the inner surfaces of the massulae and have anchor-shaped tips (Fig. 92).

Living *Azolla* contain the commensalistic blue green algae, *Anabaena azollae*, in a cavity at the base of the upper (dorsal) lobe. This cavity also occurs in the fossil (Fig. 77) with *Anabaena*-like cells preserved in the pinna lobe (Fig. 78).

Although the fossil is similar in gross overall form to living *Azolla*, living *Azolla* contains only bi-lobed pinna. Due to the general form of the fossil not being consistent with the modern identification of the genus *Azolla*, this fossil will be given a new genus name. This fossil is considered a direct evolutionary link to *Azolla*. This Azollaceae is presently under study.

The other fossil Azollaceae has been found in the form of sori of *Azolla conspicua* Snead (Fig. 95) on a reduced fertile axis (Figs. 93, 94). Both micro- and megaspore sori are produced. Only a single megaspore is produced per sori. Megaspores are up to 725 μm in length and oblong in shape. Megaspores contain both well-preserved external and internal anatomy (Figs. 95–99). These megaspores differ markedly from those of the previously described fossil Azollaceae.

Microspore sori are up to 1,025 μm in length and round to oval in shape (Fig. 100). Each sorus consists of up to thirty-two microsporangia, each containing a single massulae with many microspores imbedded within. Externally the massulae surface contains perinal prolongations with club-shaped ends (compare Figs. 92 and 101).

This is an important find because the micro- and megaspore sori are grouped together in a mass but are still produced in pairs and are attached to

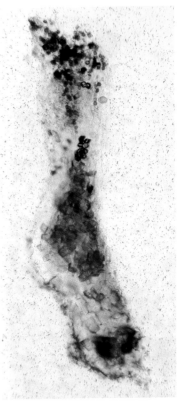

Fig. 77. Cross-section of the upper lobe in the fossil showing the cavity and heterocysts in the upper tissues.

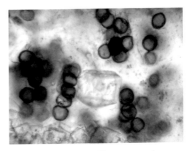

Fig. 78. Close-up of heterocysts in upper lobe.

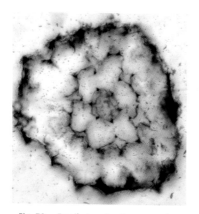

Fig. 79. Fossil pinna/root cross-section.

Fig. 80. Close-up of a fossil pinna/root vasculature.

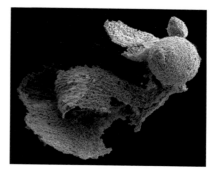

Fig. 81. Megasorus and microsorus still attached to the fossil plant.

Fig. 82. Fossil immature paired megasori.

FOSSIL PLANTS

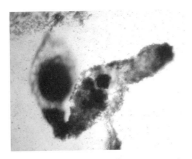

Fig. 83. Immature paired megasorus and microsorus. Focus on microsorus.

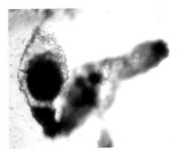

Fig. 84. Immature paired megasorus and microsorus. Focus on megasorus.

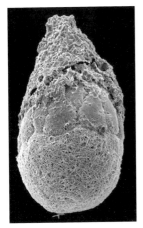

Fig. 85. Fossil megaspore with indusial covering.

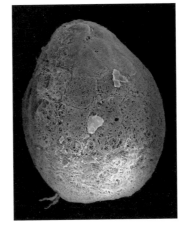

Fig. 86. Fossil megaspore.

Fig. 87. Fossil megaspore.

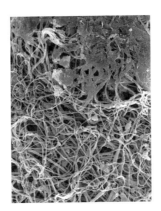

Fig. 88. Fossil megaspore perinal prolongations.

KEVIN R. AULENBACK

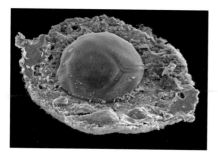

Fig. 89.   Fossil megaspore fractured obliquely.

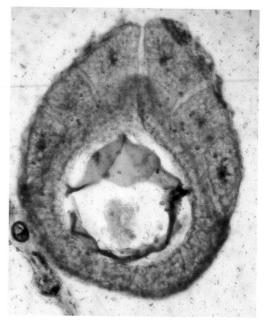

Fig. 90.   Cross-section of fossil megaspore.

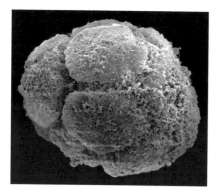

Fig. 91.   Fossil microsporangia.

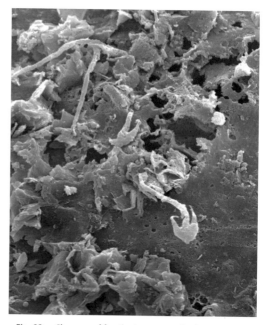

Fig. 92.   Close-up of fossil microsporangia showing anchor
shaped perinal prolongations.

an axis. The previously described fossil Azollaceae produced its spores in isolated pairs on the plant.

The types of spores associated with this mass have been related to *Hydropteris* from the St. Mary River Formation of southern Alberta (Rothwell and Stockey 1994) but the association is questioned.

The present reproductive mass remains undescribed. The plant that produced these sori would be of great interest scientifically. It would surely show a more primitive Azollaceae form.

Of the remaining two spore form genera, plant remains of *Azollopsis* (Figs. 102–104), which is considered to be the most basal of the Azollaceae, would also be of great interest. The last megaspore type, *Azolla lauta* may belong to a plant similar in form to the one presently under study. Only future discoveries will tell.

Living *Azolla* prefer warm water with acidic to neutral pH. (pH 5.5–7). They range from warm temperate to tropical areas. Spores are produced above water but fall into the water at maturity to be dispersed.

Fossil Megaspores:
Azolla fistulosa
Azolla conspicua (Parazolla)
Azolla cf. *velus*
Azolla lauta
Azollopsis tomentosa

Spermatites: Seed or Sporocarp?

Spermatites are fairly common microfossil cuticular remains found in the Horseshoe Canyon Formation. When originally described (Miner 1935), *Spermatites* were thought to have two possible origins. They were either seed cuticle remains of herbaceous monocots (Angiospermae) or the cu-

Fig. 93. *Azolla conspicua* fertile axis, top view.

Fig. 94. *Azolla conspicua* fertile axis, lateral view.

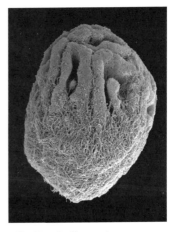

Fig. 95. *Azolla conspicua* megaspore.

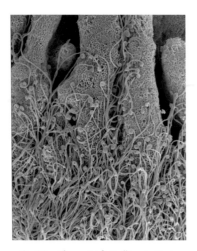

Fig. 96. Close-up of *Azolla conspicua* megaspore perinal prolongations.

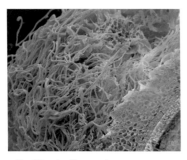

Fig. 97. *Azolla conspicua* megaspore cross-section.

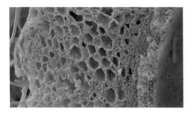

Fig. 98. *Azolla conspicua* megaspore cross-section close-up.

ticles of a sorus (Fern). Debate has not resolved their placement (Miner 1935; Binda 1968; Binda and Nambudiri 1983; Batten and Zavattieri 1995).

Spermatites is an ovate to elliptical cuticular fossil up to 1.5 mm long that was originally described as a single cuticular layer consisting of small square, hexagonal, or rectangular cells (Miner 1935) similar to Figure 105. Later remains were found to have a double layer of cuticle (Binda and Nambudiri 1983) similar to Figure 106. Most recently, *Spermatites* has been found as complete silicified remains and contain not only a double layer of cuticle but also an external coat of inflated tissue (Figs. 109–111, 113). Various immature stages have been found as well (Figs. 107, 108).

Although they were described as containing a micropyle similar to seeds (Binda and Nambudiri 1983), this is absent (Figs. 106, 107, 111, 112). Without the presence of a micropyle, the identification as a seed remain is no longer valid.

The fossils occur in high concentration with the tri-lobed pinna Azollaceae. They occur between the leaves of the plant and in the sediments, although at present none have been found attached. They are interpreted here as abortive sori of the fossil Azollaceae.

These specimens are presently under study.

Fossils named:
Spermatites transversus
Spermatites sp.

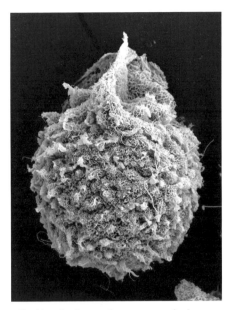

Fig. 99. *Azolla conspicua* megaspore body with massa and perinal prolongations removed.

Fig. 100. *Azolla conspicua* microsorus with multiple massulae. Each massula represent individual microsporangia.

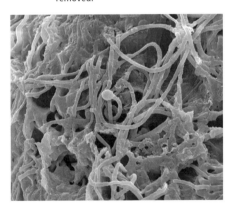

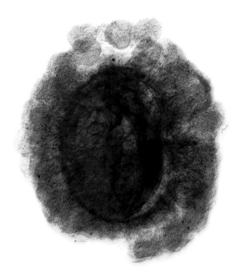

Fig. 101. *Azolla conspicua* microsporangia close-up showing club shaped perinal prolongations.

Fig. 102. *Azollopsis tomentosa* megaspore. Note that the megaspore is covered in small orbicular massae.

Fig. 103.   *Azollopsis tomentosa* microsporangia.

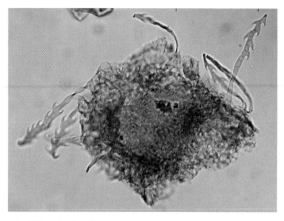

Fig. 104.   *Azollopsis tomentosa* microspore massula. Compare multibarbed perinal prolongations to prolongations in Figs. 92 and 101.

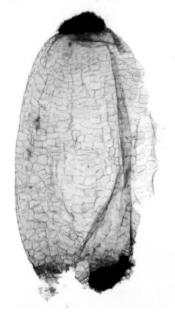

Fig. 105.   *Spermatites* sp. inner cuticle layer.

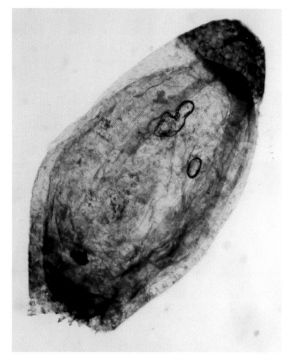

Fig. 106.   *Spermatites* sp. containing inner and outer cuticle layers.

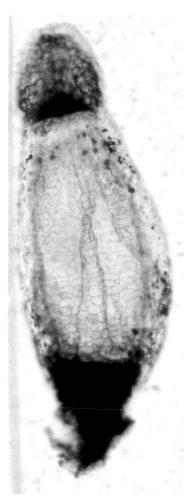

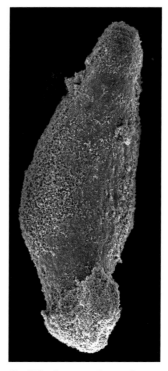

Fig. 108.   Immature *Spermatites* sp.,
external view.

Fig. 107.   Immature *Spermatites* sp.,
internal view.

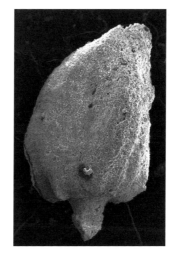

Fig. 109.   Mature *Spermatites* sp.,
external view.

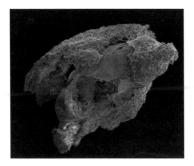

Fig. 110.   Crushed *Spermatites* sp. showing internal tissues.

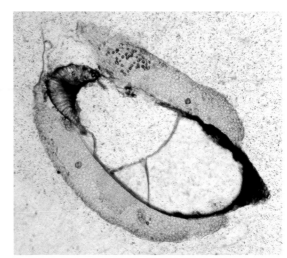

Fig. 111.   *Spermatites* sp. cross-section showing internal tissues.

Fig. 112.   *Spermatites* sp. cross-section of apex from Fig. 111.

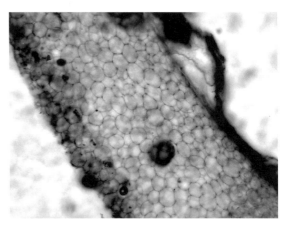

Fig. 113.   *Spermatites* sp. close-up of external tissues in Fig. 111.

SALVINIACEAE (WATER SPANGLES)

The Salviniaceae is represented by the single living genus *Salvinia* (Figs. 114, 115) with approximately ten species. *Salvinia*, like *Azolla*, is a free-floating heterosporous fern that prefers warm waters in warm temperate to tropical areas. Diagnostic of the family is the adaptation of the submerged pinna into a filiform organ from which the rupturing sori disperse the spore into the water column. This is unlike the Azollaceae that sheds its spore above water.

Living *Salvinia* is unusual in its construction in that each node of the stem produces three pinnae, two floating and one submerged. The submerged pinna is modified into branched filiform structures that mimic roots. This modified pinna is best described as a pinna/root. In the floating pinnae, the structure and distribution of papillae on the upper surface are characteristic in each species.

Unfortunately *Salvinia* is not present in the Cretaceous fossil record of Alberta and does not appear until the Early Eocene but its inclusion is important. The Cretaceous fossil floating pinnae attributed in the past to *Salvinia* are felt to belong to a highly polymorphic extinct genus (Fig. 116).

Only a single silicified fragment of floating pinna and three linear pinnae attached to a rhizome as carbon trace in siltstone have been recovered from the Horseshoe Canyon Formation representing an extinct genus. Based on this author's interpretation, whole immature plant remains have been described from the geologically equivalent St. Mary River Formation near Cardston, Alberta, as well as mature plants from Mexico.

Fig. 114. *Salvinia auriculata*, top view.

Fig. 115. *Salvinia auriculata*, lateral view.

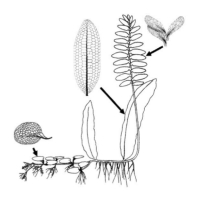

Fig. 116. Drawing of *Dorfiella* sp. plant form.

The Winding Road of Investigation

Changing Interpretations

During the processing of ironstones, a new fossil fern was found (Figs. 130–145). It is unlike any fossil fern ever discovered and represents an extinct line in the Salviniaceae.

During a literature search to find fossil comparatives and to name the fossil, it was found that the name *Hydropteris* (water fern) was already taken (Rothwell and Stockey 1994) (Figs. 117–123, 126). In the botanical code, no two wholly different plants can have the same genus name unless nested in the genus. A species in the genus must contain all the traits of the genus with the addition of species differences. The new fossil did not contain any of the generic traits of the *Hydropteris* plant already named.

Unfortunately, during this name search, it was also found that the name *Hydropteris* was previously used many years ago as a genus name for spores from the Oligocene of western Russia (Kondinskaya 1966). This discovery now relegates the fossil *Hydropteris pinnata* Rothwell and Stockey, to a homonym as the first named spore has priority.

This does not diminish the fact that the *Hydropteris* plant was thought to be a pivotal marker in the evolution of the Marsileales/Salviniales (Rothwell and Stockey 1994; Bateman 1996; Pryer 1999). Or is it? Since the new extinct fossil (Figs. 130–145) was being described, it needed to be compared against other heterosporous fossils of similar age.

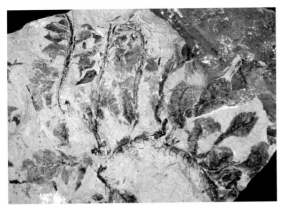

Fig. 117.   "*Hydropteris pinnata*" St. Mary River Formation, Alberta.

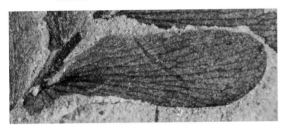

Fig. 118.   "*Hydropteris pinnata*" individual pinna from compound rachis. St. Mary River Formation, Alberta.

Fig. 119.   "*Hydropteris pinnata*" line drawing of section of compound rachis showing vasculature of the pinnae. Based on tracing from original photographs.

Fig. 120.   "*Hydropteris pinnata*" putative sporocarp (pinna). St. Mary River Formation, Alberta.

Fig. 121. "*Hydropteris pinnata*" putative sporocarp (pinna) of St. Mary River Formation, Alberta.

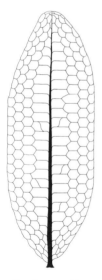

Fig. 122. Line drawing of "*Hydropteris pinnata*" simple pinna ("sporocarp").

During the investigation of "*Hydropteris pinnata*," it was noted that the specimens in the matrix did not match the reproductive nor vegetative interpretive description. Sporocarps were said to be stalked (Rothwell and Stockey 1994, Fig. 17). Areas where spores were recovered were much smaller than the interpreted sporocarps and were found behind or underneath the rhizomes as clumps (Rothwell and Stockey 1994, Figs. 9, 10).

Organs interpreted to be sporocarps, stalked and very large compared to those sampled, at various stages of development, were all devoid of spores, save those behind or underneath the rhizome. The actual spores sampled originally had depth to them as would any sporocarp or partial sporocarp clump. All the other putative sporocarps were flattened, thin, carbon traces, sometimes as slab and counter-slab, as one would expect from a pinna and not a three-dimensional organ (Figs. 120–122, 126).

The interpretation previously given of the sporocarp morphology also did not conform to those found in the ironstones from Drumheller (Figs. 93, 94), which are Azollaceae-like. The spores (*Azolla conspicua*) are identical in both, except preservational mode, compression vs. three-dimensional (Compare Figs. 95, 96, 100 to Rothwell and Stockey 1994; Figs. 18, 20, 30–32, 39).

The conclusion for "*Hydropteris*," based on actual visual inspection, is that its putative sporocarps are pinnae (Fig. 122) and the Azollaceae sori (*A. conspicua* form) are unrelated.

It was also seen that the majority of the crosiers were in a state of unfurling, which would indicate that the specimens are immature forms (Figs. 117, 118, 120, 126). Further investigations of the literature for comparatives continued.

Dorfiella auriculata Weber from the Cretaceous of Mexico was also investigated as it was originally interpreted as heterosporous in nature (Weber 1976). Amazingly, this fossil has the same dimorphic pinnae as those of *"Hydropteris."* The only difference being their fully unfurled mature form and subsequent larger length (Fig. 125).

"Hydropteris" is, in this interpretation, a *Dorfiella*, which was described many years earlier. Now there are both immature (Figs. 117, 124) and mature (Fig. 125) forms present in this dimorphic vegetative plant. Is that all?

Investigations of another plant also proved unusual. *Salvinia coahuilensis* Weber was also present with the *D. auriculata* from Mexico, both preserved on the same blocks (Weber 1976).

This presented an even better interpretation. Another even smaller new pinna form was evident in *"Hydropteris."* The smaller pinnae were always preserved in an unusual bi-fold method (wedge shaped) (Fig. 123), whereas the larger still immature single pinna exhibited characteristic unfurling of a crosier (Figs. 120, 126).

This was a third pinna type associated with the immature *"Hydropteris."* What is unusual about this small bi-fold pinna is that its only analogy in heterosporous ferns occurs in modern *Salvinia*.

Smaller bi-fold pinnae do not occur in *Dorfiella*, yet *S. coahuilensis* occurs on the same blocks in intimate association (Weber 1973, 1976) (Fig. 125). Weber (1973) even commented on the autochthonous nature of the flora.

The *S. coahuilensis* pinnae associated with *D. auriculata* are interpreted to be the same bi-fold pinnae as those from *"Hydropteris,"* only now mature (Fig. 128). It is interesting also that no immature or new growth areas were noted in *S. coahui-*

Fig. 123. *"Hydropteris pinnata"* bi-fold pinnae. St. Mary River Formation, Alberta.

Fig. 124. *"Hydroperis pinnata"* reconstruction based on specimens viewed by the author.

Fig. 125. Tracing of *D. auriculata* type specimen from Weber (1976), illustrating the position of the various pinnae types. Green = *Salvinia*-like, brown = pinnate pinnae, blue = linear pinna.

lensis. The new plant form of *Dorfiella* ("*Hydropteris*") is polymorphic in its vegetation (Fig. 116).

There are reproductive sori attached to the combined *Salvinia/Dorfiella* (Fig. 127). Unfortunately requests to process spores with new techniques did not yield any samples for processing, which is sometimes the case. Spores of Cretaceous proto-*Salvinia* are currently unknown, and it appears it will remain that way for the present.

How can we justify this reinterpretation of *Dorfiella* ("*Hydropteris*") with *Salvinia* besides that of the extremely close association of pinna types? Again we go to the fossil record to hopefully better solidify the interpretation.

Another unusual Salvinia, S. reussii Bůžek et al. (1971), is known from the Tertiary (upper Aquitanian to lower Burdigalian) of Bohemia. It contains submerged pinnae, which were interpreted as floats (Bůžek et al. 1971) (Fig. 129). Tracings of the venation of these pinnae show amazing similarity to those of the linear pinna in "Hydropteris" and Dorfiella (Fig. 122).

These floats are interpreted as vestigial pinnae as the plant went from a terrestrial habitat releasing its spores in the water (*Dorfiella*, "*Hydropteris*") to partially aquatic (Tertiary *S. reussii*), and now fully aquatic (extant *Salvinia*).

Extant *Salvinia* contains microscopic hairs on the underside of the floating pinnae, on the rachis and modified pinna/root (Fig. 115). These are used for absorption of nutrients from the water.

The Tertiary *S. reussii* contains hairs as well, but their coverage on the underside of the floating pinna is much reduced. They do cover the "floats" (submerged pinnae) and the modified pinna/root but were not found on the rachis (Bůžek et al. 1971).

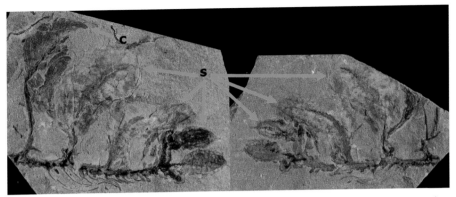

Fig. 126. "*Hydropteris pinnata*" slab and counter-slab showing the unfurling of compound pinnae (c) and single pinna (s) attached to the same rhizome. St. Mary River Formation, Alberta.

Fig. 127. Tracing of *Salvinia coahuilensis* holotype specimen, from original photograph. Weber (1973). Green = pinnae, pink = sori.

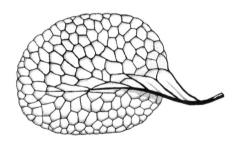

Fig. 128. Drawing of *Salvinia coahuilensis* floating pinna showing vasculature based on published specimen. Weber (1973).

Fig. 129. Reconstruction of Tertiary *Salvinia reussii* based on published figures from Bůžek et al. (1971) and adapted from Collinson (1991). Leaf at arrow showing underlying venation traced from published figure. Drawing showing submerged view of floating pinnae. Papillae placed on one floating pinna, one submerged pinna and the pinna/root only.

The *S. coahuilensis* pinnae associated with *Dorfiella* do not contain hairs nor do the pinna/roots (Weber 1973). Why would this be the case? Sori, which are just as fragile, are present and attached.

The logical interpretation is that they were not present in *Dorfiella* as the reproductive structures and floating pinnae were still attached to the plant on the land, which contained a rhizome and roots for nutrient uptake. The floating *Salvinia*-like pinna and the modified pinna/roots are merely functioning for spore dispersal in the water environment.

A full evolutionary lineage in extant *Salvinia* can now be constructed (compare Figs. 115, 116 and 129). This is based on re-interpretation of both fossils and the literature.

An Extinct Salviniaceae

Another extinct fossil Salviniaceae occurs in the Horseshoe Canyon Formation (Figs. 130–145). This is a diminutive fern that, due to its small size, would most likely be overlooked by the layperson. The fern is less than 1.5 cm in total length.

The fossil represents a complete plant (Figs. 130, 133) that grew modified pinna/roots (pinna to act as roots) similar to living *Salvinia*. The entire plant represents a rachis with multiple alternate pinnae without a rhizome or roots. Unlike *Salvinia*, the fossil pinnae, due to the lack of lamina, are like skeletal leaves with only the vasculature present (Figs. 132, 134–136). The first few pinnae are modified into underwater pinna/roots, which bear amphisporangiate sori (Figs. 130, 134, 138). The sori are round to oval in shape. Unlike modern *Salvinia*, the fossil pinna/root do not contain microscopic hairs but do contain multiple dichoto-

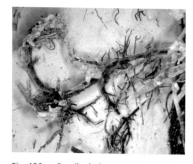

Fig. 130.   Fossil whole vegetative section.

mies of the pinna/root into finer and finer absorption structures (Figs. 130, 133, 134).

The fern contains the megaspore genera *Ghoshispora bella*, *Ghoshispora canadensis* and the microspore *Ghoshispora scollardiana*. The spores were dispersed underwater similar to living *Salvinia*.

The megaspore genus *Ghoshispora* has been used as a marker worldwide for Cretaceous outcrop in palynology. The fern became extinct in the earliest Paleocene and left no other ancestors. It is unlike any fern presently alive.

This fern is presently under study.

Fossil spore:
Ghoshispora bella
Ghoshispora dettmannii
Ghoshispora canadensis
Ghoshispora kondinskayae
Ghoshispora major
Ghoshispora rara
Ghoshispora rigida
Ghoshispora scollardiana

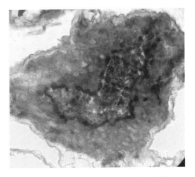

Fig. 131.   Fossil cross-section of rachis showing stele form.

Fig. 132.   Fossil frond section in rock.

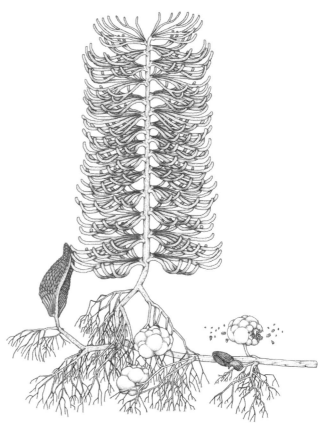

Fig. 133.   Drawing of the fossil in life position.

Fig. 134.   Fossil exhibiting rupturing sori
(middle), rachis with pinna
(top) and pinna/root (lower).

Fig. 135.   Fossil pinnae on rachis.

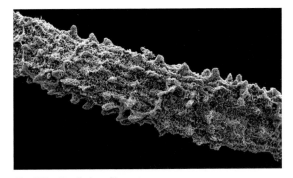

Fig. 136.   Fossil pinna filament.

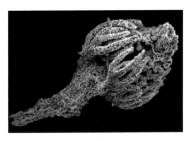

Fig. 137.   Fossil Immature crozier.

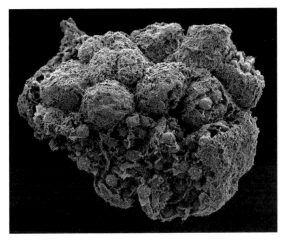

Fig. 138.   Fossil amphisporangiate sorus with mega- and
microsporangia.

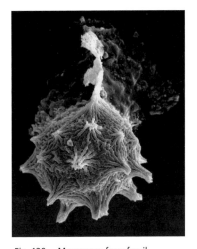

Fig. 139.   Megaspore from fossil.

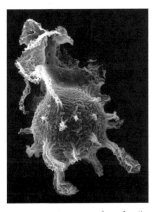

Fig. 140.   Megaspore from fossil.

Fig. 141.   Megaspore from fossil.

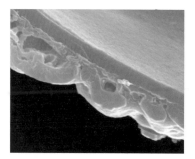

Fig. 142.   Megaspore wall in cross-section.

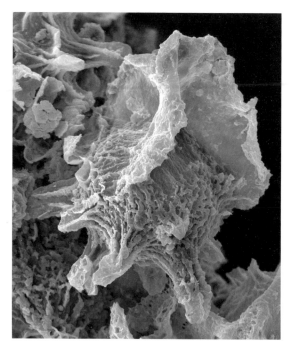

Fig. 143.   Microspore from fossil.

Fig. 144.   Microspore from fossil.

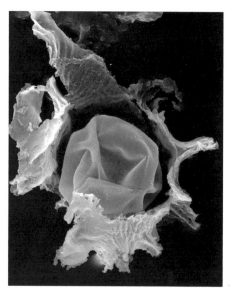

Fig. 145.   Microspore from fossil with exine surface exposed.

Marsileales

MARSILEACEAE (WATER CLOVERS)

The Marsileales are distantly related to the Salviniales (Fig. 153). The Marsileales contain the family Marsileaceae with three living genera; *Marsilea* (Fig. 146) with up to fifty species, *Regnellidium* with one species, and *Pilularia* with about five species. Like the Salviniales, these ferns are also heterosporous; however, they contain a rooted rhizome and foliage attached to long stipes. Each genus has distinctive micro- and mega-spores, sporocarp, rhizome, and pinna/stipe construction.

The family is represented in Cretaceous strata based on isolated mega- and microspores, isolated sporocarps, and isolated foliage (Chitaley and Paradkar 1972, 1973; Kovach and Batten 1989; Skog and Dilcher 1992; Lupia et al. 2000; Yamada and Kato 2002; Nagalingum 2007). Only one genus and two species of megaspore are described from the Horseshoe Canyon Formation. Recently, intact associated mega- and microspores have been recovered in an eroded sporocarp (Figs. 147–151). These remains are presently being described.

The fossil consists of a sporocarp without external covering (Fig. 147). There are eight amphisporangiate sori per side with 0–3 megasporangia per sorus. Megaspores are of the *Molaspora reticulata* form, elongate oval to spherical in shape and up to 400 µm long, 300 µm wide (Fig. 148). Microspores are of the *Gabonisporis bacaricumulus* form, up to 12 µm in diameter and spherical to round in shape (Fig. 150).

The fossil is felt to represent an early form of Marsileaceae due to morphological similarities of the sporocarp. Previously, microspores of *Gabonisporis* (Fig. 152) were thought to belong in the

Fig. 146.　*Marsilea* sp.

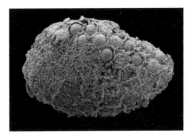

Fig. 147.　Fossil sporocarp.

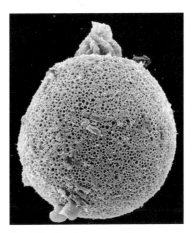

Fig. 148.　Fossil megaspore from sporocarp.

Ophioglossaceae but are here placed in the Marsileaceae due to the association of *M. reticulata* and *G. bacaricumulus*.

Living *Marsilea* prefer muddy pond or lake edges with or without water level fluctuations. *Marsilea* are warm temperate to tropical in distribution.

Fossil spore:
Molaspora lobata
Molaspora reticulata
Gabonisporis bacaricumulus
Gabonisporis labyrinthus

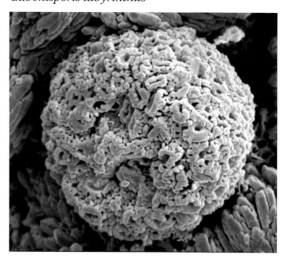

Fig. 150. Fossil microspore from sporocarp.

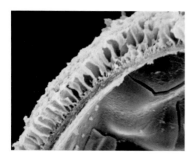

Fig. 149. Fossil megaspore wall in cross-section.

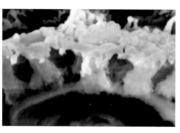

Fig. 151. Fossil microspore wall in cross-section.

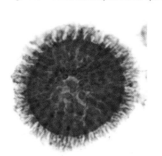

Fig. 152. *Gabonisporis bacaricumulus.*

Relationships in the Extant Heterosporous Ferns

Debate still lingers on the relationships in the extant heterosporous ferns. Are they monophyletic or polyphyletic? Presently, many authors have assumed a monophyletic origin for the Salviniales/Marsileales (Hasebe et al. 1995; Bateman 1996; Pryer 1999; Pryer et al. 2004). Based on investigations of Cretaceous fossils placed in the Salviniales/Marsileales, this appears true, although only at the progenitor ordinal level. A reinterpretation of relationships of the Marsileales/Salviniales is given in Figure 153.

The Salviniales and Marsileales are monophyletic but are distantly related in time. Both are derived from the same order of the Filicopsida, most likely the Schizaeales. The progenitor Marsileales appears to have diverged approximately 17 million years earlier from this order prior to the divergence of the proto-Salviniales. Neither gives rise to the other and they developed independently. The reproductive structures are not homologous and developed independently as aquatic dispersal methods.

The Marsileales contain a rooted rhizome and sori produced in a modified pinna (sporocarp). The sporocarp is produced above water but only releases its spore upon submersion. This modified pinna has been traditionally called a *sporocarp* and should remain named so.

Both the Azollaceae and the Salviniaceae are closely related and appear monophyletic from within the Salviniales. Morphologically both lack rhizomes and sporocarps. Although historically the reproductive structures were named sporocarps these are only sori, based on morphology. The Salviniales contain sori and pinna modified to act as roots or pinna/roots. *Azolla* subsection *Azolla* contains only a single bi-lobed floating pinna and a single pinna/root. In the subsection *Rhizosperma, A. nilotica* contains fascicled pinna/roots. During the reproductive phase in *Azolla*, sori are produced attached to vasculature stalks on the ventral (lower) pinna lobe with a modified covering lobe (involucre; Strasburger 1873 "involucrum"; Campbell 1893; Rao 1935; Konar and Kapoor 1972) produced on the dorsal (upper) lobe. The *Azolla* pinna/root does not produce reproductive structures.

The fossil Azollaceae (Figs. 71–92) contains a single tri-lobed pinna with sori attached to vasculature produced on the ventral (lower) lobe, a single modified pinna/root, and is interpreted to disperse its spore above the water, which then fall into the water column. It can be directly linked to modern *Azolla*.

Salvinia contains two non-lobed floating pinnae and a highly bifurcate pinna/root. Sori are produced underwater directly attached to the pinna/root. *Azolla* disperses its spore above the water and *Salvinia* in the water.

There has been sporadic evolution in the Salviniales with the last divergence being *Salvinia* from the newly interpreted *Dorfiella*. Previous authors have speculated on *Salvinia*'s lack of fossil Cretaceous history and have even suggested an extended ghost lineage (Pryer 1999).

Sporadic extinctions are also present such as the new Salviniaceae fossil (Figs. 130–145). This fossil also lacks a rhizome and contains the modified pinna/roots and releases its spore in the water similar to *Salvinia*. It does, however, contain a modified pinnate frond and pinnae without lamina.

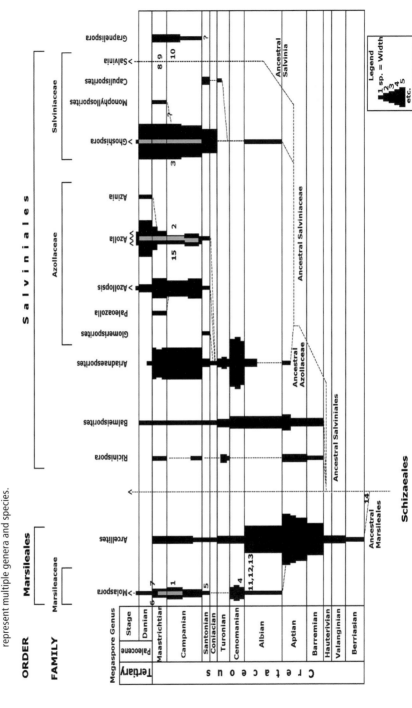

Fig. 153. Spindle diagram flow chart of suspected evolutionary relationships between Cretaceous megaspore genera based on morphology, known stratigraphic ranges and fossil mega-plant records. Black = confirmed species, grey = spore species contained by mega-plants from the Horseshoe Canyon Formation, "∧" = species continuation further than the Danian. Numerals = known mega-plant fossil species; 1. = Marsileaceae from the Horseshoe Canyon Formation; 2. = Azollaceae plant from the Horseshoe Canyon Formation; 3. = Extinct Salviniaceae from the Horseshoe Canyon Formation; 4. = Marsileaceaephyllum johnhallii; 5. = Regnellidium upatoiensis; 6. = Rodeites polycarpa; 7. = Rodeites dakshini; 8. = Dorfiella (Hydropteris) pinnata; 9. = Dorfiella coahuilensis; 10. = Dorfiella sp. Horseshoe Canyon Formation; 11. = Marsileaceaephyllum lobatum; 12. = Marsileaceaephyllum sp. B; 13. = Marsileaceaephyllum sp. C; 14. = Regnellites nagashimae; 15. = Azolla conspicua spore form from the Horseshoe Canyon Formation. The Schizaeales represent multiple genera and species.

KEVIN R. AULENBACK

75

Fig. 154.   *Osmunda regalis* reproductive and sterile fronds.

Fig. 155.   *Osmunda regalis* extracted rhizome.

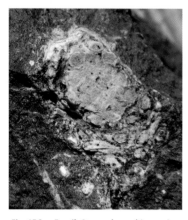

Fig. 156.   Fossil *Osmunda* sp. rhizome in ironstone.

Osmundales

OSMUNDACEAE

Osmunda

The living Osmundaceae consists of three genera, *Leptopteris*, *Osmunda*, and *Todea*. The Osmundales is an ancient order whose roots can be traced back to the Late Permian (Miller 1971). *Osmunda* (Figs. 154, 155) is well-represented in the Horseshoe Canyon by silicified rhizomes.

Dispersed spores (Fig. 163) of the *Osmunda*-type are quite common throughout the formation, but as yet identifiable intact sporangia or frond material has not been described.

Fossil rhizomes are found preserved in ironstones (Fig. 156) as well as roll out silicifications (Figs. 157, 171). When extracted from ironstones, they are clothed in the remnants of leaf bases (Fig. 159). Silicified rhizomes are up to 6 cm long and 1.5 cm in diameter. These rhizomes sometimes show bifurcation of the axis (Figs. 159, 162).

Internally the anatomy is well-preserved (Figs. 158, 160–162) and is similar to extant *Osmunda*.

Recent *Osmunda* contains a more massive rhizome and substantially larger root mass (*Osmunda* fibre) that forms a mantle around the rhizome. It is only with some physical effort that the rhizome itself can be extracted from these large masses. The fossil rhizomes are found as isolated individuals in ironstones free of *Osmunda* root fibre.

Although presently named after the living *Osmunda cinnamomea*, the fossil lacks any foliage or reproductive organs (Serbet and Rothwell 1999). It has been suggested that the species has remained unchanged for sixty-nine or so million years (Serbet and Rothwell 1999), but this must be borne out by discoveries of more complete plant parts.

Foliage and reproductive organs are diagnostic for the living ferns. Changes in pinna form or sori disposition may have occurred and cannot be discounted. Although the rhizomes appear internally similar to the modern species, the external form appears suspect. The lack of appreciable root fibre masses and smaller, uniform rhizome diameters may be indicative of species differences and casts doubt on the species longevity.

Osmunda cinnamomea (Cinnamon Fern) is found from North America to Central and South America as well as Asia. *O. cinnamomea* grows 60–150 cm tall in moist to wet, acidic humus-rich soils. It prefers semi-shade to shade in swamps, bogs, wet savannah, moist woodlands, and lake margins. Living *O. cinnamomea* is marcescent, and it is felt that the fossil form was marcescent as well.

Fossil spore:
Baculatisporites comaumensis
Osmundacidites wellmanii

The fossils on which the identification of *Osmunda cinnamomea* was based occurred in a single site. Subsequently, many more sites have been discovered, which apparently show various forms of preservation of the same rhizomes. The discussion that follows is based on these finds and their interpretation.

A Case of Preservational Bias: Is It One and the Same?

During the excavation of a fossil stump from a coal zone in Drumheller, large numbers of an unknown rhizome were discovered. Subsequently, these rhi-

Fig. 157. Fossil *Osmunda* sp. rhizome roll outs.

Fig. 158. Fossil *Osmunda* sp. rhizome cross-section.

Fig. 159. Fossil *Osmunda* sp. rhizome with bifurcate axis etched from rocks.

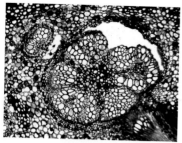

Fig. 160.   Fossil rhizome close-up of stele from Fig. 158.

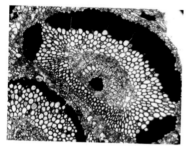

Fig. 161.   Fossil leaf departure stele close-up of from Fig. 158.

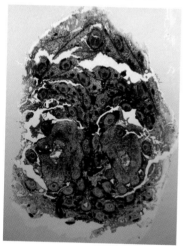

Fig. 162.   Fossil *Osmunda* sp. rhizome cross-section with bifurcate axis.

zomes have been found at a variety of sites in various coal zones.

The rhizomes are preserved in growth position as carbon/calcium phosphate traces (Figs. 164–167) in a coaly-siliceous matrix. Rhizomes are up to 30 cm long, 2 cm wide, and show bifurcations (Figs. 164, 165). When found as weathered samples, they have a characteristic form. The surfaces of the rhizomes are covered with thick and thin strands (Figs. 165–167).

On closer examination the thick and thin strands appear in sets (Figs. 166, 167), each thick strand with either one or two thin strands (one on each side of the thick strand). These strands are interpreted as the remains of the departing petiole stipe vascular trace preserved as calcium phosphate.

Internally the rhizomes contain a poorly preserved stele with surrounding ensheathing stipe vasculature (Figs. 168, 169). Individual cells are not preserved.

The forms of the rhizome are indicative of a fern and are presently interpreted as representing *Osmunda* based on the following criteria.

The silicified rhizomes named *Osmunda cinnamomea* (Serbet and Rothwell 1999) are found preserved in ironstones immediately on top of a coal seam. The rhizomes are not in growth position and consist of sections without roots or foliage. Along with them are many other plant fragments representing angiosperm and Taxodiaceae remains (Fig. 170). Their depositional environment is interpreted as rip-up lags from the underlying coal, possibly during storm events or flood stages.

In the Red Deer Valley, other rhizomes preserved in ironstones have been found, which can be identified as *Osmunda* and contain various forms of degradation in preservation (Figs. 171–

175). Some contain crushed axes with partial preservation of the outer ensheathing stipes by iron carbonates (Figs. 173, 174). Still others show outer ensheathing stipes, with the stipe vasculature preserved as carbon or calcium phosphate in silicate (Figs. 171, 172).

Some specimens preserve the main vascular strand and two lateral vascular strands in the stipe or just the main vascular strand without stipular wings (Fig. 175). These rhizomes all lack any appreciable root mass.

All rhizome forms appear to conform to the description of the original fossil rhizomes and are found in mudstones in direct contact with various coal seams. None are considered to be in growth position and occur in similar lags like the originally described rhizomes.

In the carbon trace forms found in the coals, the surface strands are identical in placement and size to those of the fossil *Osmunda* rhizomes stipe vascular traces. The fossil *Osmunda* rhizomes contain a well-preserved central trace and commonly one or two lateral traces in the stipular wings (Figs. 158, 159, 162, 173).

The central stele of the silicified rhizome is similar in form and size to those of the carbon-trace forms (compare Figs. 160 and 174 to Fig. 169). A gradation of rhizome preservational forms, from silicified to carbon trace, are present.

The carbon trace forms also preserve small masses of roots sporadically along their length (Fig. 176).

Are the carbon trace fern rhizomes the same species as the silicified rhizomes?

One could argue that both forms of rhizomes represent totally different ferns species due to the sediments each is preserved in and their preservational form.

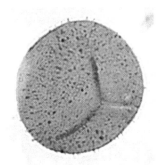

Fig. 163.   *Baculatisporites comaumensis* spore.

Fig. 164.   Freshly split slab with rhizomes. Large rhizome showing bifurcation (arrow).

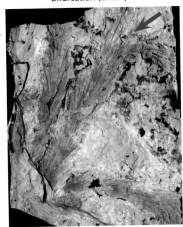

Fig. 165.   Naturally weathered slab with rhizome. Arrow indicates bifurcation.

Fig. 166. Fossil rhizome.

Are fossil species separated based on their preservational form and are there enough similarities in overall form to decide conclusively that these are the same species? If each type of fern were found separately, without intermediate forms, they would most likely be assigned different genus names. However, based on the depositional connection where each is found, overall form and habit, as well as interpreted internal structure, the carbon-trace rhizomes could easily be referred to the same genus and species as the silicified fossil rhizomes.

A question remains: are the fossils named conclusively the same species as the extant taxa?

Based on the additional intact *Osmunda* rhizomes and their form and structure, the conclusion would be no, the species longevity of *O. cinnamomea* is considered doubtful. It is felt that these fossils require further investigation.

Fig. 167. Fossil rhizome.

Fig. 168. Fossil rhizome cross-section.

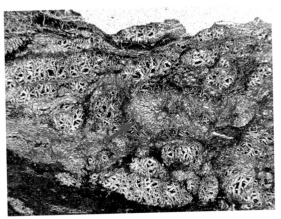

Fig. 169. Fossil rhizome stele (arrow).

Fig. 170. Turbated plant fragments in ironstone in which *Osmunda* rhizomes are found.

Fig. 171. Fossil *Osmunda* sp. rhizome roll out.

KEVIN R. AULENBACK

Fig. 172. Fossil rhizome close-up of stipe
vascular traces (median trace at
arrow) from Fig. 171 (compare
to Figs. 166, 167).

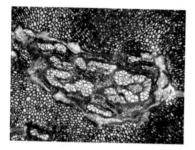

Fig. 174. Rhizome stele close-up of in
Fig. 173. Compare to Figs. 160
and 169.

Fig. 176. Associated root mat from
carbon trace rhizomes.

Fig. 173. Fossil *Osmunda* sp. rhizome cross-section containing
both silica and siderite replacement.

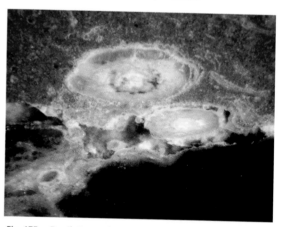

Fig. 175. Fossil *Osmunda* sp. rhizome weathered cross-section
still within matrix lacking preservation of stipular
wings and smaller associated vascular strands.

Schizaeales

SCHIZAEACEAE

The family Schizaeaceae presently contains four genera: *Anemia* (Figs. 177–179), *Lygodium*, *Mohria*, and *Schizaea*. The Schizaeaceae first occur in the late Jurassic (160 mybp) as evidenced by *Cicatricosisporites* sp. spore (Fensome 1987).

Pinnae of a fern bearing sori with spores of the *Cicatricosisporites* type (Fig. 184) have been recovered from the formation (Figs. 180–183). The pinnae are small, up to 3.75 mm long and consist of a pinna with curled edges that cover the sporangia on the lower side (Figs. 180, 181). The spores (Figs. 182, 183) are similar to spores of the extant *Anemia* in the subfamily Anemioideae. The extant *Anemia* is divided into two subgenera, the *Anemiorrhiza* and the *Anemia*, which together contain up to 117 species.

Although extremely rare as plant remains, at present it is felt that this genus may have been fairly common throughout the formation if the spore distribution is any indication. It is also felt that the rhizomes listed under the Dennstaedtiales are in fact those of *Anemia* subgenus *Anemiorrhiza*. Reasons for this possible placement are discussed under the Dennstaedtiales.

The variety of spore representing *Lygodium*, *Schizaea* (Fig. 185), and *Anemia* hint of many as yet undiscovered plant species. The fossil pinnules are as yet unstudied and un-named.

Fossil *Schizaea* and *Lygodium*-like spore:
Klukisporites foveolatus
Klukisporites notabilis
Leptolepidites tenuis

Fig. 177. *Anemia mexicana* fertile and sterile pinnae.

Fig. 178. *Anemia mexicana* sporophylls.

KEVIN R. AULENBACK

Fig. 179.   *Anemia mexicana* sporophyll
close-up.

Fossil *Anemia*-like spore:
Appendicisporites cristatus
Cicatricosisporites baconicus
Cicatricosisporites furcatus
Cicatricosisporites ornatus
Cicatricosisporites perforatus
Cicatricososporites drumhellerensis
Cicatricososporites norrisii

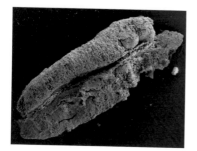

Fig. 180.   Fossil sporophyll, adaxial
surface.

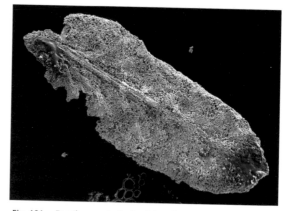

Fig. 181.   Fossil sporophyll, abaxial surface view of Fig. 180.

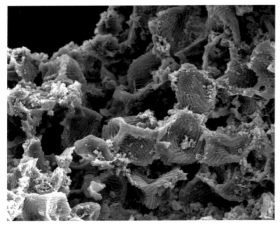

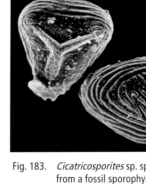

Fig. 183. *Cicatricosporites* sp. spores from a fossil sporophyll showing both proximal (upper left) and distal (lower right) polar views.

Fig. 182. Spores within a fossil sporangia.

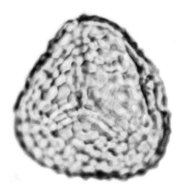

Fig. 185. *Klulisporites foveolatus* spore, polar view.

Fig. 184. *Cicatricosisporites* sp. spore, equatorial view.

Fig. 186. *Onoclea sensibilis* sterile pinnae.

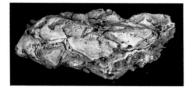
Fig. 187. Fossil *Wessiea oroszii* rhizome (*Onoclea* type).

Fig. 188. *Woodwardia virginica* reproductive pinnae.

Polypodiales

Spores representing the various families of the Polypodiales are placed together due to the (as yet) inability to key out the palynomorphs to a specific family, as many are similar in form. In the Polypodiales, there are fourteen families that have many genera that collectively contain thousands of species.

BLECHNACEAE

In the family Blechnaceae there are nine genera and up to two hundred species.

Two fern rhizomes have been found as surface collect in the formation with one representing the Blechnaceae similar to *Woodwardia*, named *Midlandia nishidae* Serbet and Rothwell (Figs. 189–192, and Serbet and Rothwell 2006, Figs. 1.1–2.4). The other, less defined as an arthyrioid Dryopteridaceae or Blechnaceae (Serbet and Rothwell 2006), is named *Wessiea oroszii* Serbet and Rothwell (Fig. 187, and Serbet and Rothwell 2006, Figs. 3.1–3.6). *Wessiea oroszii* contains the frond trace of the Onocleoid type (Ogura 1972). These two ferns were previously identified as *Blechnum* sp. A and *Blechnum* sp. B in Serbet's 1997 thesis.

These ferns represent the first occurrence of anatomically preserved Blechnoid fern's rhizomes in the fossil record. Fossil foliage of *Onoclea* is found in the Paleocene of Alberta (Rothwell and Stockey 1991), while foliage of both *Woodwardia* and *Onoclea* are known from the Paleocene of Saskatchewan (McIver and Basinger 1993).

Woodwardia are hardy temperate to subtropical ferns that prefer acidic soils (pH 5.5–6.5). *Woodwardia* contains two living species. *W. fimbriata*, also called the Giant Chain Fern, is a large deciduous fern native to China and prefers full sun to partial shade in moist soils or partial submersion in water. *W. virginica* (Fig. 188) or the Virginia Chain Fern is native to North America. It is deciduous, up to 30–60 cm tall (30 cm in partial sun to full shade) with new growth bronze-green in colour.

Onoclea sensibilis (Fig. 186) is a monotypic fern found in eastern North America and Asia.

A variety of spore types representing the Polypodiales are found in the formation (Fig. 193).

Fig. 189. Fossil *Midlandia nishidae* (*Woodwardia*-like) stipe.

Fossil Polypodiales spore:
Hazaria canadiana
Hazaria sheopiarii
Laevigatosporites albertensis
Laevigatosporites adiscordatus
Laevigatosporites discordatus
Laevigatosporites druggii
Laevigatosporites ovatus
Laevigatosporites sp.
Polypodiisporites amplus
Polypodiisporites favus

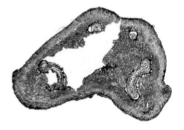

Fig. 190. Fossil *Midlandia nishidae* (*Woodwardia*-like) stipe.

DENNSTAEDTIACEAE

The Dennstaedtiaceae contains eleven fern genera. The family's earliest-known appearance as a megafossil is in the Paleocene of Saskatchewan (McIver and Basinger 1993), although spores are found well into the Upper Cretaceous (Fig. 197).

Fossil rhizomes attributed to the family (Fig. 198) and isolated stipe material (Figs. 195, 196) are found in the Horseshoe Canyon. The rhizomes were previously compared with the fern *Microlepia* (Serbet and Rothwell 2003) and are named *Microlepiopsis bramanii* Serbet and Rothwell and *Microlepiopsis aulenbackii* Serbet and Rothwell. Many more rhizomes of this type have been found since.

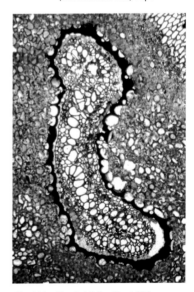

Fig. 191. Fossil *Midlandia nishidae* (*Woodwardia*-like) close-up of linear stele trace from Fig. 190.

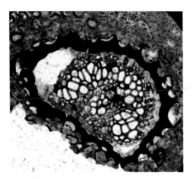

Fig. 192. Fossil *Midlandia nishidae* (*Woodwardia*-like) close-up of circular stele trace from Fig. 190.

Fig. 193. *Laevigatosporites* sp. spore.

Fig. 194. *Dennstaedtia punctiloba* sterile fronds.

Individual *Dennstaedtia*-like stipe sections are up to 20 mm long and 2.5 mm wide. Internally most of the tissues are collapsed, but the fossils do show two curved arms with remnants of the cross member, the stele being an omega shape (Figs. 195, 196). This stipe material has not been described.

Extant *Microlepia* is an evergreen fern pantropical in distribution. Extant *Dennstaedtia punctiloba* (Fig. 194) occurs in North America.

Fossil spore:
Varirugosisporites tolmanensis
Leptolepidites bullatus

Although rhizomes identified as *Microlepiopsis bramanii* and *M. aulenbackii* are named from the formation (Serbet 1997; Serbet and Rothwell 2003), they are felt by this author to possibly belong to another family.

A Case of Similarity in Form: Problems Inherent in Investigation and Comparisons

In 1989, fern rhizomes were found in growth position by the author at a site in Midland Provincial Park. Many more were etched from ironstones. The rhizomes were subsequently published as rhizomes of *Microlepiopsis bramanii* and *M. aulenbackii* (Dennstaedtiaceae), due to the similarity of these fossils to those of living *Microlepia* (Serbet 1997; Serbet and Rothwell 2003).

Although *Microlepia* and the family Dennstaedtiaceae commonly produce rhizomes similar to the fossils in morphology, they are not the only ones that do so. In cross-sections of the fossil, the vascular construction of the amphipholic solenostele (Figs. 199, 200) is found in many orders, families, and genera of ferns (Bower 1923; Ogura 1972).

Fern pinnules were in the same ironstones that produced these rhizomes. These pinnules contained spores of the *Cicatricosisporites* type, which, due to their form, are placed in the Schizaeaceae. Also in these rocks, large quantities of dispersed *Cicatricosisporites* spores were found and none were identified as belonging to *Leptolepidites* or *Varirugosisporites* of the Dennstaedtiales were encountered. These discoveries prompted further relationship enquiries.

In the Schizaeaceae genus *Anemia*, subgenus *Anemiorrhiza*, contain creeping rhizomes with pinna traces on the dorsal surface and roots on the ventral surface. Internally the rhizomes contain an amphipholic solenostele (Boodle 1901) like the fossil (compare Fig. 201 to Fig. 202).

In the fossil the cortex contains a slightly sclerotic outer half and a strongly sclerotic inner half. The stellar ring is bounded by an inner and an outer endodermis. The central ground tissue is also sclerotic. The pericycle forms a layer 2–4 cells thick on the inner and outer side, and there is an inner and outer layer of phloem, each consisting of an almost continuous layer of thick-walled sieve tubes mostly one cell thick and filled with dark contents. This is almost identical to that described by Boodle (1901) for *Anemia mexicana* (subgen. *Anemiorrhiza*).

In the fossil, a layer of parenchyma separates the phloem on both sides from the xylem. The xylem is 1–5 tracheids in thickness with both large and small tracheids scattered in the xylem. Protoxylem elements are found most commonly mesarch in placement. This differs only marginally from *A. mexicana* (Boodle 1901).

Fig. 195. Fossil *Dennstaedtia* sp. stipe.

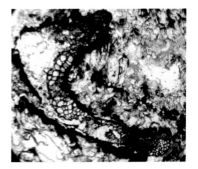

Fig. 196. Fossil *Dennstaedtia* sp. Stipe stele lateral arm.

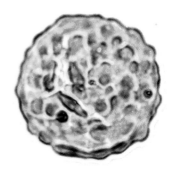

Fig. 197. *Varirugosisporites tolmanensis* spore.

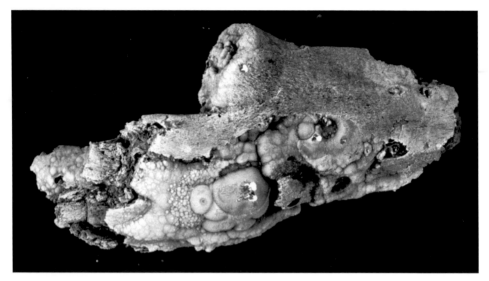

Fig. 198. Fossil *Microlepiopsis aulenbackii* rhizome.

Fig. 199. Fossil *Microlepiopsis aulenbackii* rhizome cross-section.

Fig. 200. Fossil *Microlepiopsis aulenbackii* rhizome cross-section showing stele bifurcation.

The simple stipe trace of the fossil was previously described as the *Loxsoma* type (Serbet and Rothwell 2003). This form is found in the *Adiantum*, *Dennstaedtia*, *Lindsaea*, and *Microlepia*. *Anemia* was previously described as the having a *Gleichenia* type as found in *Anemia*, *Mohria*, and *Gleichenia* (Ogura 1972). Nevertheless, in *Anemia* subgenus *Anemiorrhiza* (Fig. 204), this trace appears very similar, if not identical, as described and figured to the *Loxsoma* type (Bower 1923, 1926; Ogura 1972). This similarity of trace form was noted by Bower (1926).

The fossil has a stipe trace that is saddle-shaped, flat-topped to slightly bowed, arch shape with divergent supports that have hooked ends turned inwards similar to *Anemia phyllitidis* (Boodle 1901). Protoxylem forms the upper saddle and lower ends of the lateral arms and hooks. The lateral arms are formed of enlarged metaxylem composed of scalariform tracheids, which may extend into the hooks.

In the fossil, the phloem, which is separated from the xylem by a layer of parenchyma, forms a layer covering the outer surface of the xylem-arch and continues on the inner surface about halfway up each arm similar to *Anemia phyllitidis* (Boodle 1901). The phloem consists of sieve tubes and parenchymatous cells scattered among them, and are distinguished from them by their dark contents similar to *Anemia phyllitidis* (Boodle 1901). The phloem is thickest on the outer and inner side of the arms. In the fossil, the overall description of the stipe trace appears similar to *A. phyllitidis* (compare Figs. 203 and 204).

The stipe departure morphology in both *Anemia* and the fossil are similar. Both contain stipe departures, which are situated directly over the leaf gap. This stipe departure differs from all *Mi-*

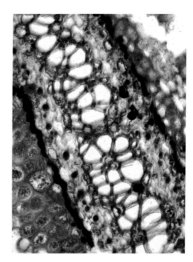

Fig. 201.   Fossil *Microlepiopsis aulenbackii* rhizome stele close-up.

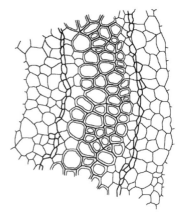

Fig. 202.   Drawing of *Anemia mexicana* stele (redrawn from Boodle 1901).

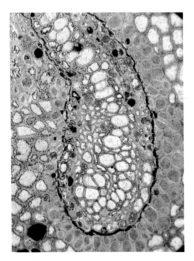

Fig. 203. Fossil *Microlepiopsis aulenbackii* rhizome departing stipe.

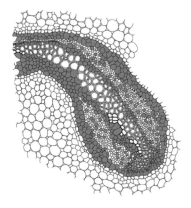

Fig. 204. Drawing of *Anemia phyllitidis* stipe (redrawn from Bower 1923).

crolepia species, which exhibit diagnostic stipe departures more or less lateral to the leaf gap (Nayar and Kaur 1963a,b). These lateral stipe departures give the rhizome and stipe of *Microlepia* a distinctive shape in cross-section with the stipe attached by one arm to the stele during departure.

The fossil root stele is asymmetrical in shape. It contains a single layer of phloem, the cells of which are distinguished by their dark contents, surrounding numerously irregularly placed xylem tracheids. A layer of parenchyma one to two cells thick separates the xylem from the phloem (Fig. 205). This is similar to the root of *A. phyllitidis* described by Boodle (1901), where the base of the root stele in the cortex is diarch, but the middle region of the xylem plate is occupied by numerous irregularly placed tracheids, giving the stele a slightly asymmetrical appearance.

The past mention of similarity in the presence of aerenchymatous tissue in both *Microlepiopsis* and *Dennstaedtiopsis aerenchymata* from the Eocene (Serbet and Rothwell 2003) is somewhat misleading. Aerenchymatous tissue is a product of habitat adaptation and not an indicator of generic relationship. This type of tissue is found in a variety of cryptogams as well as many aquatic angiosperms.

Arenchymatous tissue bands are found laterally in the stipes of *Microlepia* (Nayar and Kaur 1963a,b) but is absent in the fossil and *Anemia*.

It is also unusual that the two closely related species of *Microlepiopsis* occur exactly at the same site. Modern fern species of the same genus may overlap in range but in slightly differing habitats. If species are in close association, plant hybridization frequently occurs. Since the material from the Kent's Knoll locality is interpreted by this author as a single event and environment, it is unlikely two species were present.

The two fossil species were separated based on the subsequent trace in *M. bramanii* developing from two lateral steles with incurved arms (Fig. 206B) (Serbet and Rothwell 2003; Figs. 3D, 6B) to form a single-saddle shaped stele (Fig. 206A), whereas *M. aulenbackii* retained the saddle-shaped stele through its departure (Fig. 206A) (Serbet and Rothwell 2003; Figs. 5B,C,D,F,G, 6A). The morphology of the rhizome is such that the departing petiole trace slants away and at an upward angle from the rhizome. In larger fossil rhizomes viewed, this appears more pronounced.

In a serial cross-section cut of such a stem, the distal portions of the departing petiole would allow only for a cut across the two arms of the trace (Fig. 206D). The arms of the trace are also more developed and the comparatively softer bridge of the arms is usually recessed in the silica.

In sections previously published (Serbet and Rothwell 2003), the stele bridge trace is not present, similar to other cells absent in the area, indicating cellular collapse (Serbet and Rothwell 2003; Figs. 3D, 6B). The stipe stele trace shows severe in-folding and decay in the bridge area. Xylem and other internal stele tissue extend into the area without indications of stele walls encircling two independent steles. This would indicate the arm is missing due to decay or recessed into the trace during desiccation prior to fossilization (Fig. 206C). Irrespective of where the trace gap occurred in the rhizome, the absence of the encircling stele walls would still indicate decay of tissue in the area.

The rounded silica ends of the petiole traces and their comparatively similar height would indicate that the fern stipes contained abscission bases and was more than likely deciduous. Marcescent

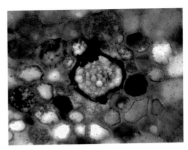

Fig. 205. Fossil *Microlepiopsis aulenbackii* rhizome departing root trace.

ferns such as *Osmunda* and non-deciduous ferns do not preserve petiole traces with rounded well-formed ends and have staggered petiole heights due to stipe decay.

Based on the above information, a strong case exists for the fern rhizomes of *Microlepiopsis bramanii* and *M. aulenbackii* belonging to a new single deciduous species of the *Anemia* subgenus *Aneimiorrhiza* and not the Dennstaedtiales. In the new genus, the species *bramanii* would have priority.

It is felt that these rhizomes require additional study into their affinities.

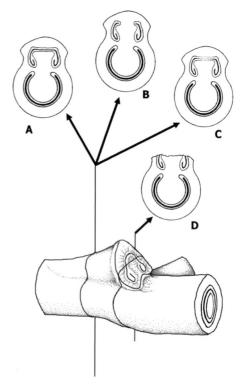

Fig. 206.   Drawing of *Microlepiopsis* sp. rhizome with cuts and possible results. A = complete arm traces; B = two separate arm traces; C = degraded arm traces; D = incomplete arm cut.

Gleicheniales

GLEICHENIACEAE

The family consists of three living genera: *Sticherus*, *Dicranopteris*, and *Gleichenia*.

Fig. 207. *Gleicheniidites senonicus* spore.

The Gleicheniales is an ancient order that first appears in the Upper Carboniferous and Permian. There are three Cretaceous genera; *Gleichenoides* from Malaysia; *Gleichenopsis* from Greenland; and *Didymosorus* from Germany. In the United States, Cretaceous fossils have been placed in the extant *Gleichenia*. The oldest confirmed Gleicheniaceae in North America is from the Raritan/Lower Magothy Formation of the Potomac Group, Turonian (90 million) of New Jersey (Gandolfo et al. 1997) named *Boodlepteris turoniana*. Presently, the fossil record from the Horseshoe Canyon Formation consists of spores (Fig. 207). The fossil plant awaits discovery.

Living *Gleichenia* is a medium-sized fern (20–300 cm tall) restricted to India, Malaysia, New Guinea, Australia, New Zealand, and Polynesia. It generally prefers moist to wet, acidic soil in full sun.

Fossil spore:
Dictyophyllidites harrisii
Gleicheniidites concavisporites
Gleicheniidites senonicus

MATONIACEAE

This family contains two genera: *Phanerosorus* and *Matonia*. This genus at one time was thought to be related to Marsileaceae. The fossil record of the family in the formation consists of miospores.

The appearance of the spores (Fig. 208) in the formation suggests they may be reworked from earlier formations.

Fossil spore:
Matonisporites phlebopteroides
Matonisporites sp.

Fig. 208. *Matonisporites* sp. spore.

Cyatheales (Tree ferns)

The Cyatheales contains eight families: the Cyatheaceae, with five genera and over six hundred species; Loxomataceae, with two genera with a single species in each; Dicksoniaceae, with one genus and thirty species; Cibotiaceae, with one genus and eleven species; Culcitaceae, with one genus and two species; Thyrsopteridaceae, with one genus species; Plagiogyriaceae, with one genus species; and Metaxyaceae, with one genus species.

Although four spore types appear in the formation (Fig. 213), mega-fossils have not been found. This is problematic due to the large size of living plants (Figs. 209–212) as well as fossil trunks found in other parts of the world. Does the spore record reflect actual tree fern spore? The present author has engaged in active searches for megaplant remains over many years in the formation without success. It is felt that if present the megaplant was either very site-specific or differing in form from those growing today. The spore may also not represent actual tree ferns but some other unknown fern form.

Cyatheales exist in wet montane or cloud forests. Tree ferns are evergreen warm temperate to tropical plants that thrive in cool moist conditions and acidic soils (Figs. 209, 210).

Fossil spore:
Cyathidites concavus
Cyathidites minor
Cyathidites punctatus
Deltoidospora sp.

Fig. 209.   *Cyathea* sp., Australia.

Fig. 210.   *Cyathea* sp. habitat, Australia.

Fig. 211.   *Cyathea* sp. tree trunk naturally buried in soil.

Fig. 212.   *Cyathea cooperi* pinnule with sporangia.

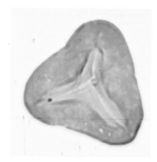

Fig. 213.   *Cyathidites minor* spore.

Ophioglossales

OPHIOGLOSSACEAE

Although the fossil microspore *Undulatisporites sp.* is placed here, it may prove to belong to another family as was the case with *Gabonisporis*, which in the past was placed here as well, but is now known to belong to the Marsileaceae. *Undulatisporites sp.* is therefore left here with reservation.

The Ophioglossaceae contains three genera: *Botrychium*, *Ophioglossum*, and *Helminthostachys*. These ferns generally prefer acidic soils and moisture in open-sunny or partially-shaded conditions and are deciduous to non-deciduous.

Fossil spore:
Undulatisporites sp.

Incertae Sedis

Fig. 214. *Liburnisporis adnacus* spore.

Spores of Unknown Affinity

The following spores have yet to be placed in existing families or genera, due mainly to lack of similarity to modern cryptogams. Two spore forms are figured to show variability (Figs. 214, 215).

Fossil spore:

Bacutriletes sp.

Biretisporites sp

Concavissimisporites variverrucatus

Divisisporites maximus

Echinatisporis levidensis

Echinatisporis spinilabia

Inundatisporis tappanii

Inundatisporis vermiculisporites

Liburnisporis adnacus

Marattisporites rousei

Monophyllosporites corynemorphus

Polycingulatisporites reduncus

Polycingulatisporites triangularis

Polycingulatisporites sp.

Radialisporis radiatus

Taurocusporites segmentatus

Trisolissporites radiatus

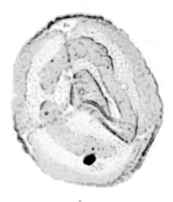

Fig. 215. *Polycingulatisporites reduncus* spore.

Unknown Remains of Filicophyte Affinity

Many fern parts are found in the formation but unfortunately not everything has been studied or published to date. These unknowns are placed here with the hope that someone in the future may take up the challenge of identifying them.

RHIZOME

This rhizome occurs at Kent's Knoll locality only. It is represented by two specimens and neither has been sectioned for internal structure. The departing trace appears fern-like (horseshoe shaped). The rhizome is stout, 9 mm long (Fig. 216), and covered by short scales on the ventral surface (Fig. 217). It is felt it may represent possibly a primitive form of the Marsileaceae as sporocarps were etched out of the blocks, which also contained these rhizomes.

NON-RHIZOME PARTS

Many stipes and crosiers are found in thin sections, but unfortunately these have not been keyed out. Some of the different types are shown.

In the stipes, some contain aerenchymatous tissues (Figs. 219, 221, 222), which indicate living in moist environments, while others do not (Figs. 220, 223, 224). Preservation may be variable as iron carbonate, silicates, or carbon trace. Most appear to be identifiable.

Large carbon-trace fern pinnae can be found on silicified coal bed slabs. Some are very common (Fig. 218), while others are rare. The more common form may belong to the carbon-trace rhizomes attributed to the Osmundales. Both can occur on the same slabs.

Three-dimensional crosiers have been extracted from iron carbonates (Fig. 225). These are small, only up to 4 mm long, and appear to contain a saddle-shaped stele.

Section of plants that resemble cryptogam gametophytes have also been recovered (Figs. 226, 227). These are formed of carbon and cuticles encased in silica and are up to 10 mm long.

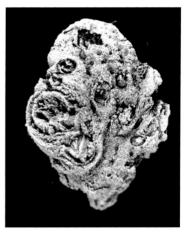

Fig. 216. Fossil rhizome, dorsal view.

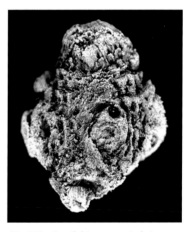

Fig. 217. Fossil rhizome, ventral view.

Fig. 218.   Common fossil fern.

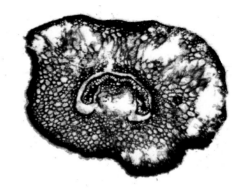

Fig. 219.   Unknown stipe.

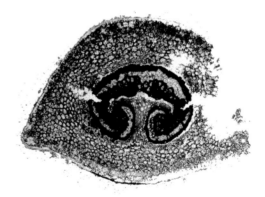

Fig. 220.   Unknown stipe.

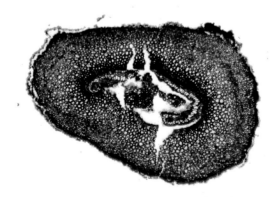

Fig. 221.   Unknown stipe.

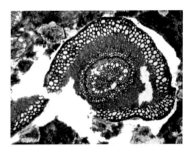

Fig. 223.   Unknown stipe.

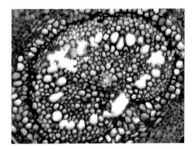

Fig. 224.   Close-up of stele in Fig. 223.

Fig. 222.   Close-up of Fig. 221.

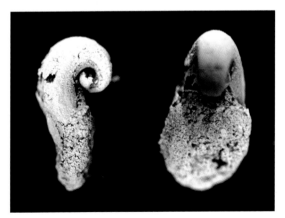

Fig. 225.   Lateral (left) and facial (right) views of immature crosier.

Fig. 226. Fossil gametophyte.

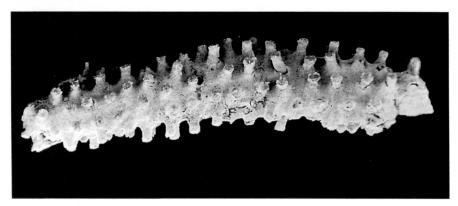

Fig. 227. Fossil gametophyte.

Cycadophyta

Presently, true Cycads (Cycadales) have not been found in the formation. The Cycadophyta in the form of *Nilsonia*-like leaves in the formation, which has been placed in the Cycadales in the past, can be argued to be a Bennettitales (false cycad).

Caytoniales

The Caytoniales is an extinct group of Mesozoic seed ferns. These seed ferns were most prominent in the Jurassic, but they have also been found in Triassic and Cretaceous sediments (Stewart 1987). Presently, only pollen represents this family in the formation (Fig. 228). If the fossil plant were ever found in the formation, it would be of great interest scientifically due to the possibility of it showing the final stage of development in the order prior to extinction.

Fossil pollen:
Vitreisporites pallidus

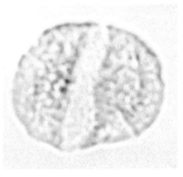

Fig. 228.   *Vitreisporites pallidus* pollen.

Bennettitales (False cycads)

Seed Cones

The Bennettitales are an extinct order of plants related to the Cycadales. In the formation many seed cones with construction reminiscent of the Williamsoniaceae   (Bennettitales/Williamsoniales) are found preserved as carbon or calcite/silica replaced remains (Figs. 229–232). Although similar to Williamsoniaceae, there are major differences, suggesting a new family should be erected for the fossils.

Fig. 229. Fossil seed cone naturally weathered.

The seed cones are oblong in shape, rounded basally, and range in size up to 5.0 cm long and 2.0 cm wide, and they are composed of tightly packed helically disposed sporophylls best identified as cupules (Figs. 232–234). Distal cupules are abortive/sterile forming a compact apical mass (Figs. 230, 231).

The individual cupules are long-stalked and distally form a peltate, octagonal to pentagonal face. Each face contains a prominent ovule chamber cavity opening situated in the upper middle portion. Cupules are tightly packed against adjacent cupules (Fig. 235).

The main axis of the cone is poorly defined and composed of many loosely assembled vascular strands. These vascular strands are embedded in parenchyma tissue and arise basally without any indication of a peduncle or base.

Upon entering a cupule, the single vascular strand bifurcates repeatedly in the tissue surrounding the point of ovule attachment and ovule chamber cavity (Figs. 233, 234); the strands terminate in the cupule distal face.

The point of ovule attachment is round to oval with a long wide-collared ovule chamber cavity. The ovule chamber cavity flares distally (Figs. 229, 231, 234).

The long, wide ovule chamber cavity is easily discernible in weathered or prepared specimens (Figs. 229, 231–234). It is formed of a thin cuticular layer lining the cavity and multiple layers of encircling cells that terminate proximally and delineate the point of ovule attachment and the ovule chamber cavity (Figs. 232, 238).

The point of ovule attachment and ovule chamber cavity lengths are highly variable. Longer ovule chamber cavities make the point of ovule attachment elliptical, while a shorter ovule chamber cavity makes the point of ovule attachment appear round.

Only seedless ovulate cones have been identified in the Horseshoe Canyon collections. However, a mass of loosely dispersed cupules, considered
to represent a single disaggregated cone, from the
Oldman Formation has a similar cupule as those
from the Horseshoe Canyon. These weathered cupules (Figs. 236) contain the ovule, embryo, ovule
chamber wall, and remnants of the longitudinally
oriented cells, parenchyma tissue, and external
cupule distal face.

In specimens from the Oldman Formation,
only a single ovule is produced per cupule. The
ovule is oval to elliptical in shape and fills the
ovule chamber cavity. The ovule is capped by the
cupule distal face. The face bears a much-reduced
ovule chamber cavity opening situated in the upper middle portion that is similar to specimens
from Drumheller which have seeds shed.

Dispersed seeds and ovules prepared from
some of the Oldman Formation cupules contain
a diminutive linear embryo situated in a lateral
aspect near the apex (Fig. 237). The apex of the
seed contains a low, offset, crescent-shaped ridge
opposite the embryo.

The cones from the Horseshoe Canyon Formation are found associated with the leaves of *Nilsonia* that are presently attributed to them. These
cones, which are morphology similar in form to
those of the Bennettitales/Williamsoniales (Fig.
239), are presently being studied. It is hoped that
future discoveries of siliceous cones will allow
viewing of the external seed cone face.

Fossil pollen:
Cycadopites carpentieri
Cycadopites follicularis
Cycadopites nitidus

Fig. 230. Large fossil seed cone freshly exposed.

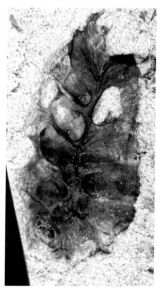

Fig. 231. Small seed cone partially weathered.

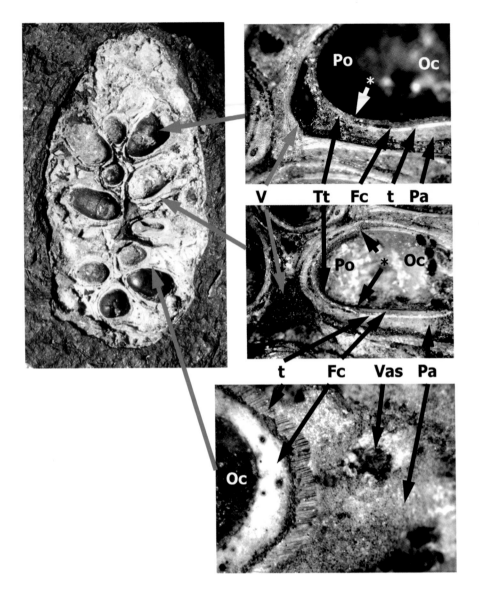

Fig. 232. Naturally weathered, calcite permineralized seed cone (left). *Upper right:* close-up of cupule showing a proximal end, longitudinal section of the point of ovule attachment and the mud in-filled ovule chamber cavity. *Middle right:* close-up of cupule showing a proximal end, longitudinal section. *Lower right:* close-up of cupule cross-section and the upper portion of the ovule chamber cavity. Fc = fleshy cells lining ovule chamber cavity; Oc = ovule chamber cavity; Pa = parenchyma; Po = point of ovule attachment; t = tubular layer; Tt = transfusion tissue; v = vascular bundle of stalk; Vas = vascular tissues; * = denotes end of transfusion tissue and beginning of fleshy cells lining the ovule chamber cavity.

Fig. 233. Chemically altered basal portion of a seed cone showing vasculature of the cupules.

Fig. 234. Close-up of vascular strands surrounding the cupule in Fig. 233.

Fig. 235. Tangential section of naturally weathered subsurface view of immature seed cone.

Fig. 236.   Eroded cupule. Oldman
Formation, Alberta.

Fig. 237.   Apical view of Seed from cupule. Arrow points to
embryo. Oldman Formation, Alberta.

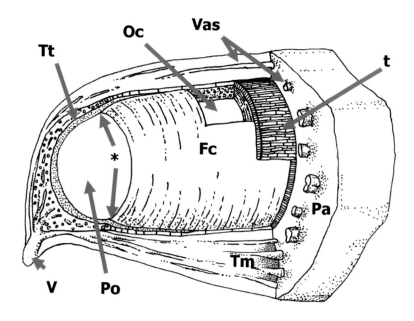

Fig. 238.   Cut away line drawing of mature cupule from the Horseshoe Canyon showing construction with tissue
types: Fc = fleshy cells lining the ovule chamber cavity; Oc = ovule chamber cavity; Pa = parenchyma;
Po = point of ovule attachment; t = tubular cells; Tm = tissue mass with surrounding vasculature; Tt =
transfusion tissue of ovule seat; V = vasculature bundle of stalk; Vas = vasculature strands; * = denotes
end of transfusion tissue and beginning of fleshy cells lining the ovule chamber cavity.

110

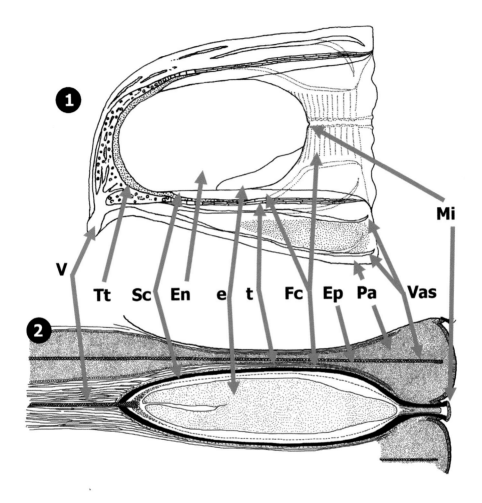

Fig. 239. Comparison of tissues in Horseshoe Canyon seed cupule and *Bennettites* cupule. Arrows denote tissue types. 1 = reconstruction of longitudinal section of Horseshoe Canyon cupule with semi-mature ovule. Stippling in micropylar area represents immature closed position of tissues prior to maturation and seed expulsion. 2 = *Bennettites albianus* cupule with apically fused interseminal scale (redrawn from Stopes 1918); e = embryo; En = endosperm; Ep = epidermis of interseminal scale; Fc = fleshy cells lining the ovule chamber cavity (deliquescent layer of Stopes 1918); Mi = micropyle; Pa = parenchyma; Sc = seed coat; Tt = transfusion tissue; t = tubular cells; v = vascular bundle of stalk; Vas = Vascular strand.

Is what you see what you've got?

The interpretation of two-dimensional fossils in the third-dimension

When investigating plant finds, one of the most important questions for proper identification should be "What would this plant look like if I could actually hold it in my hand?" This is easiest for individual leaf specimens, but it becomes more difficult with leaf disposition on a branch or fruit or seed cone. The three-dimensional disposition and shape of these organs is important because it allows for an understanding of morphology, which in turn allows for a better understanding of relationships with other taxa. Here are three examples:

1. Are the leaves of the fossil branch held in opposition or are they alternate? Do they spiral up the branch or are they decussate? This can be ascertained by preparation of the stalk or stem by removal of sections of carbon trace. If already missing, one should look for the tell-tail signs of leaf or branch scars. *Parataxodium* (see Taxodioideae, Sequoieae; *Parataxodium*) as a carbon trace is a good example; although it seems to bear its leaves in opposition, on closer examination they are alternate and helically arranged (Figs. 240, 241). This becomes more difficult with fruit or cones.

2. During investigations of the conifer *Mesocyparis* (see Cupressaceae), it was seen that the three-dimensional seed-cones of *Mesocyparis umbonata* McIver and Aulenback were held in opposite decussate fashion up the branch (Fig. 242). During the viewing of comparative material, it was

Fig. 240. Drawing of carbon trace *Parataxodium* sp. branch with apparent opposite phyllotaxy.

Fig. 241. Close-up of lower area in Fig. 240. Carbon trace removed revealing leaf bases of as yet un-exposed leaves (arrows) showing helical phyllotaxy.

found that the Paleocene fossil *Mesocyparis borea-lis* McIver and Basinger was described as having only opposite seed cones (McIver and Basinger 1987) but was in fact opposite decussate as well (Fig. 243). The three-dimensional *M. umbonata* indicated a need for re-evaluation of *M. borealis*. The evolutionary relationship between the species is now much stronger than previously thought.

3. While investigating Bennettitalean seed cones from the formation, other comparative materials were looked for. Seed cones of *Microzamia gibba* (Reuss) Corda from Europe (Kvaček 1997) were compared to them. Although *M. gibba*'s physical weathered trace structure appeared comparable (Figs. 244; 1–3, 2a, 3a), its actual interpretive description was not. It had previously been described as Cycadales-like (Kvaček 1997) but, based on actual morphology, it is not (Fig. 244; A–C).

The seed cones of specimens from Alberta and *Microzamia* were previously described as similar with bracts and paired ovules (Serbet 1997). Bracts do not exist in either seed cone genera. The structures present in the fossils are those of ovule infills (cupules). How do we know this?

This is most easily shown in Figure 244 and is most easily demonstrated by taking a cylinder of wood that represents the cone of *Microzamia* or the Drumheller fossil. Drill holes around the cylinder to mimic the cavities left by the expelled seeds. Paint latex over half the wooden cylinder, which represents the sediments being laid down. When dried, peel the latex off. The latex peel is now an accurate example of the fossil cupulate Bennettitalean seed cones before us.

As for a bract ovule example, take an extant seed cone of *Sequoiadendron* or any *Zamia* and let it dry to the point of having the bracts shrink. Force latex into the semi-dried cone representing

Fig. 242. *Mesocyparis umbonata* Immature seed cones.

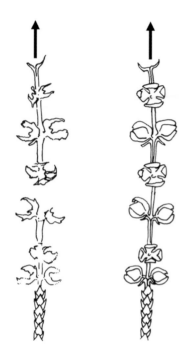

Fig. 243. *Mesocyparis borealis* seed-cone axis line drawings. *Left:* drawing of carbon trace, USMN 10222. *Right:* line reconstruction.

the burial sediments. When dried, peel the latex off. This mould looks totally dissimilar to the two cupulate Bennettitalean fossil seed cones (Fig. 244). This test shows well the actual form of the seed cone trace.

These problems are not isolated and are encountered as new fossils, views, concepts, and interpretations come into play. What you see is not necessarily what you have and sometimes specimens must be sacrificed to give as great a three-dimensional view as possible. Only through investigation and interpretation, and re-investigation and re-interpretation, does the science move on, and we come closer to what is actually in our hands, to see the plant as it was living, not as a fossil.

Nilsonia-like leaves

Nilsonia is an organ genus erected for fossil linear leaves with pinnae attached to the top of the midrib, pinnae being differently segmented to divided irregularly, thin cuticle and haplocheilic stomata, and assigned to the Cycadales (Harris 1961). *Nilsonia*-like leaves are a common occurrence in the formation and are preserved as carbon trace, calcite or silica. Leaves are sometimes found in site association with the previously discussed Bennettitales seed cones.

Previously two *Nilsonia* leaf types from the formation were described based on the degree of dissection of the leaf blade (Bell 1949). It can be seen that there is a high degree of variability of dissection from entire to highly dissect even in individual leaves (Figs. 245–247). It is most likely that there is only one species present in the flora with highly variable leaf forms.

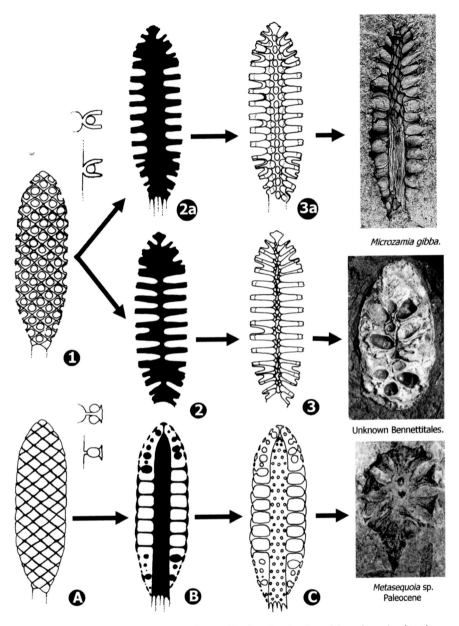

Fig. 244. Ovule in-fills (cupules) vs. bract ovules. Working from line drawings of three-dimensional seed cones of the Bennettitales from Drumheller and *Microzamia gibba* from Europe as well as an example of a seed cone bearing bracts (*Parataxodium* sp. or Cycadales), we go from plant to carbon trace fossil to fossil mould with trace removed. Numerals 1, 2, 3 = Drumheller specimens without defined axis; 1, 2a, 3a = *Microzamia gibba*, stalked form; A, B, C = generic seed cone with bracts.

KEVIN R. AULENBACK

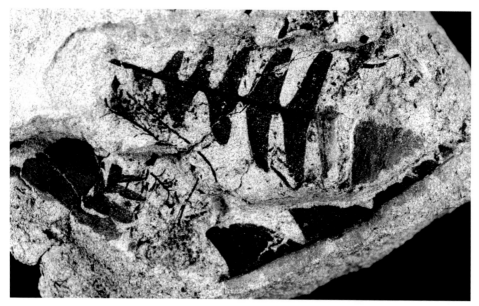

Fig. 245. *Nilsonia*-like leaves.

Fig. 246. *Nilsonia*-like leaves.

Leaves in the formation are strap-like, up to 15 cm long and 4.3 cm wide, and characterized by a prominent rachis, which is concealed by the overlying pinnae. Petioles are very short.

The pinnae have rounded lobes, which are never toothed. In many specimens the surface laminae appears corrugated due to the underlying vascular strands. Vascular strands are parallel without bifurcations.

Stomata are restricted to the ventral side of the pinnae in between the vascular strands. After chemical preparation of better preserved specimens, numerous hairs or trichomes were found on the ventral side as well. Stomatal morphology, although very possible to view, has unfortunately not been investigated in these leaves. This is problematic as leaves of *Anomozamites* or *Pterophyllum* (Bennettitales; syndetocheilic stomata) are very similar in gross morphology and are sometimes mistaken for *Nilsonia* (Cycadales; haplocheilic stomata). It is only by epidermal characters that these can be distinguished from *Nilsonia*.

In cross-section the rachis appears to have contained five vascular strands, a large basal median, and two smaller lateral strands on each side. More lateral strands may have been present as Figure 251 may represent an upper segment of leaf. The pinnae can be seen as attached to the top of the rachis with a tissue flange supporting the pinnae from below (Fig. 251).

The pinnae show a very weak palisade layer on the dorsal aspect covered by a small-celled epidermis (Fig. 250). Ventral epidermal cells are slightly inflated and interrupted by stomata. The bulk of the mesophyll contains unspecialized cells. Sclerenchyma groupings of three to eight cells are found on the ventral and dorsal aspect of the vascular strands in the mesophyll.

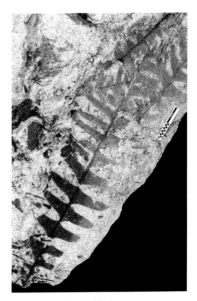

Fig. 247. *Nilsonia*-like leaves.

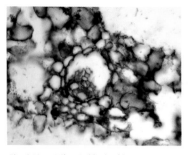

Fig. 248. *Nilsonia*-like leaf laminae vascular strand.

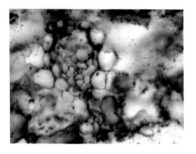

Fig. 249. *Nilsonia*-like leaf laminae vascular strand.

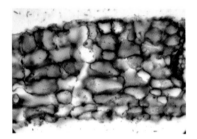

Fig. 250. *Nilsonia*-like leaf laminae mesophyll between vascular strands.

The vascular strands contain a bundle sheath one cell thick. The xylem consists of up to nine cells and the phloem is poorly preserved (Figs. 248, 249). Some pinnae show the presence of a resin canal situated towards the outer edge of the blade. The canal contains a distinct epithelial lining. Resin bodies are absent between the vascular strands.

These leaves remain enigmatic as to thier placement in the Cycadophyta. Although placed in *Nilsonia*, stomatal configuration may place this leaf form elsewhere. Until confirmed as *Nilsonia* by stomatal configuration, these organs should be referred to as *Nilsonia*-like. Their association with the Bennettitales seed cones and the lack of any Cycadales seed cones recovered from the formation casts doubt on their generic identification. If found to be a *Nilsonia* species in the future, then a reassessment of this and other seed cone/leaf associations in other formations may be in order.

It is of note that the seed cones described as *Microzamia* (Kvaček 1997), which bear striking physical similarities to those of the Horseshoe Canyon Formation (Fig. 244), contain leaf imprints similar to *Nilsonia* or *Nilssoniopteris* on the same samples, although attributed to the leaf form *Jirusia jirusii* of the Cycadales (Kvaček 1997), which are not site associated.

In the Horseshoe Canyon Formation, it is uncertain if the pollen data reflect true *Cycad* pollen or that of a *Cycad*-like representative. Presently the pollen form genus *Cycadopites* may represent both the Cycadophyta and *Ginkgo* in the formation (Fig. 252).

This *Nilsonia*-like leaf form appears to have become extinct just prior to or during the Cretaceous/Tertiary event. The last occurrence in the Cretaceous of Alberta is from the Scollard Formation with the discovery of Bennettitales seed cones and associated *Nilsonia*-like leaves.

The plant form from the Horseshoe Canyon is speculated to have been a large herb-like or small shrub-like plant growing with its leaves in a shuttlecock or rosette form similar in appearance to the Birds Nest fern, *Asplenium nidus.*

Fig. 251. *Nilsonia*-like cross-section of partial stipe and leaf blade attachment.

Fig. 252. *Cycadopites fragilis* pollen.

Fig. 253.   Silicified stem.

Fig. 254.   Cross-section of Fig. 253.

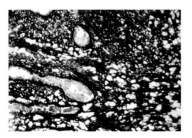

Fig. 255.   Close-up of cross-section of Fig.
254.

Plants of Suspected Cycadophyta Affinity

Some unusual plant remains in the form of stem and possible reproductive remains are suspected to represent the Cycadophyta.

A stem was found at the site, which produced the silicified *Nilsonia* leaves. This stem was short with annulated constrictions (Fig. 253). Internally it contains large resin-cell areas and highly reduced vasculature (Figs. 254, 255).

Stems have been found at the sites that produce the carbon-trace seed cones of the Cycadophyta. These stems also have annulated constrictions but lack detail that would show structure (Figs. 256–260).

A single structure consisting of multiple leaf-like appendages (sporophylls?) was also recovered. It is felt that this may possibly represent a fragment of a pollen cone (Figs. 261, 262).

Fig. 256.   Carbon trace stem.

Fig. 257.   Carbon trace stem
            chemically altered.

Fig. 258.   Carbon trace stem
            chemically altered.

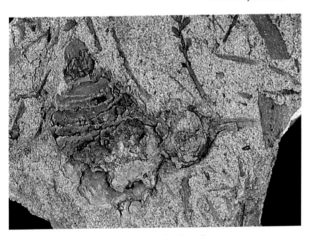

Fig. 259.   Carbon trace stem chemically altered.

Fig. 260.   Close-up of stem from Fig. 259.

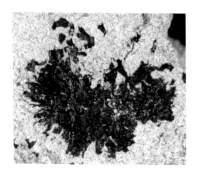

Fig. 261.   Suspected partial pollen cone.

Fig. 262.   Highlighted counter specimen of Fig. 261.

Ginkgophyta

Ginkgoales

GINKGOACEAE (MAIDENHAIR TREE)

Ginkgo

Only *Ginkgo biloba*, from Chekiang, western Hubei and possibly Zhejiang Provinces in China exists today, but many species are recognized from the fossil record around the world. *Ginkgo* first appears in the Jurassic (Tralau 1968), where it becomes species-rich and extends to the present.

Known in Chinese as the Peh-K'o tree, *Ginkgo* is well known from the Horseshoe Canyon. It occurs at many plant localities from impressions in ironstone, siltstone, and sandstone. Fan shape leaves with the dichotomous linear venation are indicative of *Ginkgo* (Figs. 263–265). Fossil leaves from the formation were named *Ginkgoites* sp. (Bell 1949).

Leaves are variable in size, with variable dissection of the lamina from whole to highly dissect. Leaves sometimes appear corrugated due to the underlying vascular strands. These strands bifurcate and fan out from the petiole.

Some sites have produced fossils so well preserved that the leaves can be literally pulled off the sediments. This is due to the thick resistant cuticles on both sides of the leaf.

Although *Ginkgo* leaves are quite common, silicified seeds appear rare. The fossil seeds, as preserved sclerotesta, attributed to the genus are oval with a lateral keel (Fig. 266) and are comparable to those of recent *Ginkgo* (Fig. 267), although much smaller overall (up to 1.1 cm wide and 1.4 cm long). The sclerotesta is up to 0.5 mm thick. Another

Fig. 263. *Ginkgo biloba.*

Fig. 264. Fossil *Ginkgo* sp. leaf in sandstone.

Fig. 265.   Fossil *Ginkgo* sp. leaf in ironstone.

Fig. 266.   Fossil *Ginkgo* sp. seed chemically extracted from rock. Only the sclerotesta is preserved.

even rarer occurrence is seeds which preserve the sarcotesta as well as the stony sclerotesta.

The sarcotesta is unlike that found in modern *Ginkgo*. The sarcotesta contains a lateral keel in line with the underlying stony sclerotesta lateral keel (Fig. 268). The sarcotesta is thickest (up to 2.75 mm) in the middle and proximal end of the seed and thins towards the distal end (0.5 mm thick). The proximal end of the seed sarcotesta contains an apex attachment point that is circular (3.75 mm in diameter) and forms a collar-like or flared flat-topped apex of the seed. The distal end contains a reduced collar, which is cone-shaped. The sarcotesta of the fossil is much thinner when compared to modern *Ginkgo* (Fig. 269). These seeds have not been formally described.

Ginkgo pollen has, in the past, been grouped with the Cycadophyta. A partial *Ginkgo* pollen cone fragment that contained pollen is figured in Serbet 1997 (unpublished thesis, Figs. 36, 37). *Zonosulcites parvus* (Fig. 270), which was at one time placed in the Nymphaeales, is presently thought to belong to *Ginkgo*.

Recent *Ginkgo biloba* is considered to be a warm temperate to subtropical tree. *Ginkgo* is dioecious with sexes on separate trees. Female trees bear fruit that produce a foul odour upon rotting. *Ginkgo* grows up to forty metres tall with a spread of up to seven metres wide, although they are slow growers (30 cm per year) under ideal conditions. *Ginkgo* is deciduous with vivid golden fall foliage.

Ginkgo can be grown as far north as Edmonton in the province of Alberta. A few small living examples can even be seen in yards around Drumheller.

Fossil pollen:
Cycadopites carpentieri
Cycadopites follicularis
Cycadopites nitidus
Zonosulcites parvus

Fig. 267. Fossil *Ginkgo* sp. seed still partially in rock and modern *Ginkgo biloba* seed for comparison. Note similarity of sclerotesta form and partial preservation of sarcotesta preserved as silica infill in rock and surrounding the sclerotesta.

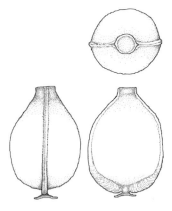

Fig. 268. Three illustrated views of a Horseshoe Canyon Formation fossil *Ginkgo* sp. seed external sarcotesta. *Upper right:* apical view. *Lower left:* lateral view. *Lower right:* facial view.

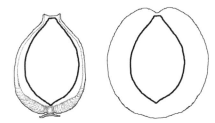

Fig. 269. Line drawings of Horseshoe Canyon fossil (left) and modern (right) *Ginkgo* seeds showing the outer sarcotesta thickness and shape in relationship to the inner sclerotesta.

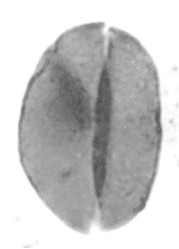

Fig. 270. *Zonosulcites parvus* pollen.

Fig. 271. *Pinus parviflora*, japanese white pine with pollen cones.

Fig. 272. *Abies*-like seed cone scale ventral view. Dinosaur Park Formation. The impressions left by the two seeds can still be seen (arrows).

Coniferophyta

Although the Mesozoic Era is sometimes considered as the age of Cycads, in the Horseshoe Canyon Formation it would be better termed the age of conifers, in particular the Taxodiaceae. It is in the Upper Cretaceous that the Taxodiaceae (Redwoods) reach their greatest diversity; by the end of the Cretaceous they decline.

CONIFEROPSIDA

Coniferae

Some conifers such as *Sequoia*, *Metasequoia*, *Glyptostrobus*, or *Araucaria*, which have been commonly thought of as Cretaceous Alberta trees, are not present in the Horseshoe Canyon flora. This misconception is mainly due to the use of leaf-form genera that has been a common practice in the past. Presently this classification has fallen into disuse and for many reasons should be avoided.

When viewed with the naked eye, many unrelated families may have overall leaf features that place them in common form genera (i.e., *Araucaria*, *Cunninghamia*, *Taiwania*, and *Athrotaxis* foliage). It is only through microscopic investigation of external and internal morphology that they can be identified correctly.

PINACEAE (PINES)

The family Pinaceae contains three subfamilies: Abietoideae, Laricoideae, and Pinoideae. The Abietoideae contains six genera: *Abies*, *Keteleeria*, *Cathaya*, *Pseudotsuga*, *Tsuga*, and *Picea*. The Laricoideae contains three genera: *Pseudolarix*, *Larix*, and *Cedrus*; and the Pinoideae contain one genus, *Pinus* (Fig. 271).

The first occurrences of pines similar to those that exist today are from the Upper Cretaceous, although seed cones identified as Pinaceae appear as far back as the Triassic (Stewart 1987).

In Alberta, *Abies*-like seed cone scales have been found in ironstones from the Dinosaur Park Formation (Figs. 272, 273). The cone scales consist of a bract which has the impression areas for two seeds (Fig. 272, arrows). The bract contains numerous resin canals in the tissue. It is hoped that one day this type of cone scale will be found in the Drumheller area.

Abies are large coniferous, monoecious evergreens native to cooler northern temperate zones. *Abies* contains approximately forty recent species.

Even older yet, a single undescribed Pinaceae seed cone, 2.7 cm long, has been recovered from the Oldman Formation (Fig. 274). Its bracts are shingle-like, and, although weathered, it appears similar to those of *Picea*.

Picea albertensis Penhallow (1908) from the Horseshoe Canyon Formation was never figured and, unfortunately, could not be located. It is unknown if it has been verified by any subsequent authors, but it appears to be a valid species.

Individual *Picea*-like seed-cone scales found in the formation have been tentatively placed in *Picea* and represent an extinct new species. The bracts are found both silicified and as carbon trace. The seed-cone scales are large, up to 4.0 cm long by 2.5 cm wide in carbon-trace specimens (Figs. 275, 276). A smaller (3 cm long) silicified bract has been previously illustrated (Serbet 1997). Based on these bracts, the cone that contained them must have been exceptionally large and would be well worth looking for in the fossil record.

As well, in the formation isolated *Picea*-like needles have been found (Figs. 277, 280, 281). The

Fig. 273.   *Abies*-like seed cone scale dorsal view. Dinosaur Park Formation. Counter piece of Fig. 272.

Fig. 274.   Pinaceae seed cone from the Oldman Formation.

Fig. 275. Partially weathered carbon/calcium phosphate trace *Picea* sp. bract.

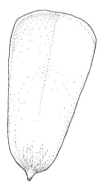

Fig. 276. Drawing of fossil *Picea* sp. bract.

Fig. 277. Fossil *Picea*-like needle abaxial surface showing stomatal bands.

needles were apparently long and slender with an internal structure similar to that of *Picea* based on fragments, up to 2.1 cm long, collected from dissolved ironstones. The fossil consists of an epidermis with a single row of epidermal cells, a fibro-vascular bundle, and two resin canals in the mesophyll tissue that are short and punctuated.

There are up to fifty living species of *Picea*. *Picea* are monoecious, coniferous evergreen trees native to the cooler parts of the northern hemisphere.

Pinus-like seed cones have also been found. Seed cones are up to 3.5 cm in diameter with bracts spirally disposed (Fig. 282). The bract tip displays a thickened apex and reduced umbo. Two small inflated oval seeds 2.0 mm long with outwardly facing micropyle are attached to the dorsal aspect of each bract. The cones, with seeds still intact, are considered slightly immature based on similar dispersed seeds up to 3.0 mm long found at other sites.

A wedge shaped needle has also been found (Figs. 278, 279). The wedge shape would indicate that the needle was part of a needle fascicle, which would indicate the needle is *Pinus*-like, although its vascular structure (Fig. 279) does not vary appreciably from the *Picea*-like form.

It is unknown what other genera of the Pinaceae may be represented in the flora as many palynomorphs appear present (Fig. 283). Only time will tell.

Fossil pollen:
Alisporites bilateralis
Alisporites grandis
Cedripites cretaceous
Cedripites sp.
Parvisaccites sp.
Pityosporites alatipollenites
Pityosporites constrictus

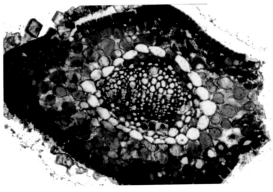

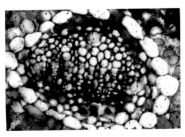

Fig. 279. Close-up of vascular cylinder from Fig. 278.

Fig. 278. Fossil *Pinus*-like needle cross-section.

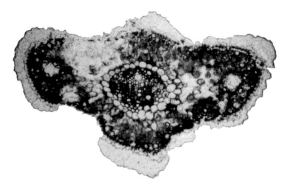

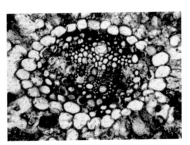

Fig. 281. Close-up of vascular cylinder from Fig. 280.

Fig. 280. Fossil *Picea*-like needle cross-section.

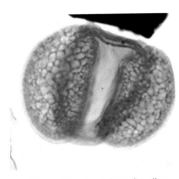

Fig. 283. *Alisporites bilateralis* pollen.

Fig. 282. *Pinus*-like seed cone cross-section.

TAXODIACEAE (REDWOODS)

The earliest representatives of the Taxodiaceae are structurally preserved seed cones from the Jurassic (Jongmans and Dijkstra 1972; Yao et al. 1998). By the end of the Cretaceous, many of the genera alive today were present.

With the use of Taxodiaceous mega-fossil remains, related genera or species are readily identifiable in the fossil record. This allows for a better understanding of the palaeoecology and climatology of the era when reconstructing the palaeohabitat. Some give a clearer picture as to environments while others give startling relationships.

The family Taxodiaceae contains one genus, *Metasequoia*, which has been commonly called a "living fossil." It will be seen that the family as a whole is more deserving of this claim.

There are nine living genera presently in the Taxodiaceae, these are: *Sequoia, Metasequoia, Sequoiadendron, Cunninghamia, Athrotaxis, Cryptomeria, Taxodium, Glyptostrobus,* and *Taiwania*. Of these up to four living genera, *Cunninghamia, Taxodium, Taiwania,* and possibly *Athrotaxis* are found in the formation. One fossil genus related to *Sequoiadendron, Parataxodium (Drumhellera),* is also found in the formation with the possibility of even more genera to be added with future discoveries.

Previously, many attempts have been made using cladistics to either subdivide the Taxodiaceae or merge the Taxodiaceae with the Cupressaceae (Endlicher 1847; Eckenwalder 1976; Liu and Su 1983; Hart 1987; Price and Lowenstein 1989; Brunsfeld et al. 1994; Kusumi et al. 2000). Although a merger with the Cupressaceae appears natural, the subdivision of the Taxodiaceae is always controversial. Even the use of rbcL gene sequences, although insightful, did not resolve known or suspected clades such as the *Cryptomeria, Glyptostrobus,* and *Taxodium* clade (Brunsfeld et al. 1994).

The pollen record in the formation also falls far short of identifying individual genera in the Taxodiaceae or for that matter the Taxodiaceae, Cupressaceae, or Taxaceae combined. Under the light microscope the pollen grains appear similar. They are small, round, and sometimes apically papillate. Even with living genera, only with the use of TEM (Transmission Electron Microscope) and SEM (Scanning Electron Microscope) do the pollen grains show diagnostic surface or internal structure that can be used to separate them at the generic level.

In the fossil record, only if pollen is found in intact identifiable pollen cones is it possible to separate the dispersed pollen. However, even in pollen cones of living genera size ranges may vary in different trees of the same species; or even different pollen cones from the same tree may produce slightly different grain sizes due to stress or cone maturity. With this knowledge, it is understandable why only three fossil pollen form genera are known that represent up to eight plant genera from three families.

For reasons of text clarity and better understanding, the Taxodiaceae will be divided into the following subfamilies and tribes as follows:

Subfamily Taxodioideae
1. Tribe Sequoieae
Sequoiadendron, Parataxodium, Metasequoia, Sequoia and *Athrotaxis.*
2. Tribe Cryptomerieae
Cryptomeria, Taxodium and *Glyptostrobus*
3. Tribe Cunninghamieae
Cunninghamia and *Taiwania*

Fig. 284. *Parataxodium* sp. fractured cross-section of silicified seed cone containing seeds.

Fig. 285. *Parataxodium* sp. fractured cross-section of calcite replaced seed cone containing seeds.

Taxodioideae

SEQUOIEAE

Parataxodium Arnold and Lowther 1955, *a deciduous Sequoiadendron*

This is one of the most abundant conifers in the Horseshoe Canyon Formation and the most mis-identified and mis-interpreted. Leaves, seed cones, and pollen cones are abundant (Figs. 284–310) in a variety of lithologies from mudstones to sandstones as silicified (Figs. 284, 304–306), calcified (Figs. 285, 303), carbon-trace (Figs. 286, 302) or natural moulds (Figs. 287, 288).

Horseshoe Canyon specimens show branches up to 6 cm long with linear leaves up to 2 cm long and 2 mm wide, disposed in a sub-opposite to alternate pattern in a single plane (Figs. 289, 291). Conversely, in immature three-dimensional specimens, the leaves are disposed in a helical fashion (Figs. 290, 292). In all specimens stomata are longitudinally disposed in two bands on the dorsal side of the leaf (Figs. 291, 293, 294) and orientated parallel to the axis of the leaf with straight-walled epidermal cells. Internally the leaves contain a single vascular bundle with a single resin canal. Persistent bud scales are found on long shoots and are absent on short lateral shoots (Figs. 289, 290).

Seed cones are situated on long stalks up to 4.5 cm long and are composed of helically disposed peltate bracts with up to thirteen inverted ovules in two rows per bract (Fig. 301). Seed cones can range from 1.1 cm to 5.5 cm in length. The seeds have two lateral wings (Fig. 307).

Pollen cones are panicle-like (Figs. 296, 310), composed of helically disposed bracts and situated alternate to pseudo-opposite, each in the axils of a linear leaf similar in morphology to vegetative leaves. Pollen cone axes contain short and long shoots similar to leafy axes. Pollen cone sporophylls contain two microsporophylls rarely three in a single row (Fig. 298). Pollen is round with an apical papillae (Figs. 299, 300).

Seed cones with stalks, pollen cones panicles, as well as long and short shoot vegetative axes were deciduous similar to extant *Metasequoia*.

Previously, all specimens of this *Sequoiaden-dron*-like form from the Horseshoe Canyon were erroneously placed in the following:

Sequoiites dakotensis (Brown) Bell 1949, seed cones.
Juniperites gracilis Bell 1949, pollen cone branch.
Drumhellera kurmanniae Serbet and Stockey 1991, pollen cone branch.
Elatocladus intermedius (Hollick) Bell 1949, foliage held in single plane.
Sequoiites artus Bell 1949, foliage in helical disposition.

In 1955 a conifer from Alaska was described as *Parataxodium wigginsii* Arnold and Lowther (Fig. 311). This conifer was thought to be an intermediary between *Taxodium* and *Metasequoia* (Arnold and Lowther 1955). Investigations of specimens of *Parataxodium wigginsii* from the type locality of Alaska by this author show that they are very similar to specimens from Drumheller, including stomatal orientation and distribution.

In the Horseshoe Canyon Formation pollen cones attached to branches from sectioned ironstones were published as *Drumhellera kurmanniae* (Serbet and Stockey 1991). Identical pollen cones were later etched out of the ironstones and found

Fig. 286. *Parataxodium* sp. fractured cross-section of carbon trace seed cone without seeds.

Fig. 287. *Parataxodium* sp. mould roll out of sideritized seed cone.

Fig. 288. *Parataxodium* sp. calcite replaced and imprint branches.

Fig. 289. *Parataxodium* sp. extracted silicified leaves.

to be identical to those of silicified *Parataxodium*-like seed cones and foliage from the formation based on stomatal configuration of the leaves.

Based on the author's investigations of both Alaskan and Drumheller specimens, and the botanical rules of priority, all specimens of *Drumhellera* should be assigned to *Parataxodium* and, only due to minor differences in pollen grain size, the species name *kurmanniae* should be retained.

Parataxodium is a member of the redwoods that presently includes *Sequoiadendron*, *Sequoia*, *Metasequoia*, and *Athrotaxis* (Figs. 312–322).

Although in the past the fossil foliage has been called *Sequoia*-like, it is only distantly related. Living *Sequoia sempervirens* is a polyploid in its chromosome count, and it has been speculated that it may represent a cross between *Sequoiadendron* and a tree similar to *Metasequoia* or a *Metasequoia*-like ancestor (Stebbins 1948; Khoshoo 1959). This deciduous *Sequoiadendron* (*Parataxodium*) may represent a more plausible predecessor. *Metasequoia* does not make an appearance in the fossil record of Alberta until the early Paleocene, based on leaf epidermal and seed-cone structure.

Metasequoia, although called a living fossil, is actually a highly evolved deciduous *Sequoiadendron* directly related to *Parataxodium*. Contrary to LePage et al. (2004), the past putative identifications of Cretaceous *Metasequoia* (Hollick 1930; Baikovskaya 1956; Samylina 1962, 1964; Bell 1963; Lebedev 1976, 1979, 1982, 1987; Patton and Moll-Stalcup, 2000) are here considered erroneous and are presently interpreted to represent *Parataxodium*. *Metasequoia*'s first occurrence in the fossil record is in the Paleocene, and it is presently recognized by three species in the Tertiary: *M. occidentalis* (Chaney 1950), *M. milleri* (Rothwell and Basinger 1979), and *M. foxii* (Stockey et al. 2001).

Extant *Metasequoia* is found in south-central China.

Sequoiadendron giganteum, the Sierra Redwood, is the tallest tree on Earth with specimens up to a hundred metres tall and a basal trunk diameter of seven metres. It is a monoecious, evergreen, warm temperate conifer that prefers full sun and deep, cool, well-drained, acidic to neutral soils (pH 5.5–6.5). It is frost resistant but drought tender.

Fossil pollen:
Sequoiapollenites papillapollenites
Sequoiapollenites paleocenicus
Taxodiaceaepollenites hiatus

Fig. 290. *Parataxodium* sp. silicified leaves showing helical phyllotaxy.

Fig. 291. *Parataxodium* sp. mature leaves held in a single plane.

Fig. 292. *Parataxodium* sp. Immature leaves helically arranged.

Fig. 293. *Parataxodium* sp. stomatal band.

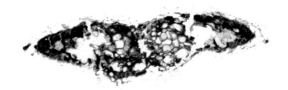

Fig. 294. *Parataxodium* sp. leaf cross-section.

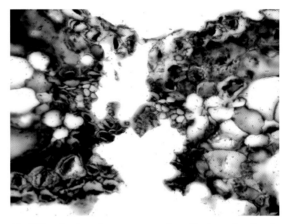

Fig. 295. *Parataxodium* sp. cross-section of leaf vascular strand.

Fig. 296. *Parataxodium* sp. pollen cone branch.

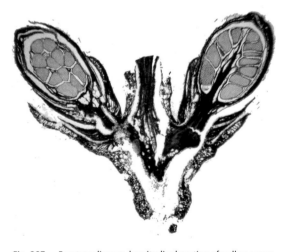

Fig. 297. *Parataxodium* sp. longitudinal section of pollen cones.

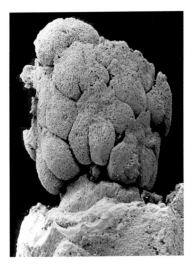

Fig. 298.   *Parataxodium* sp. pollen cone with covering bracts removed. Note 2–3 pollen sacs per sporophyll.

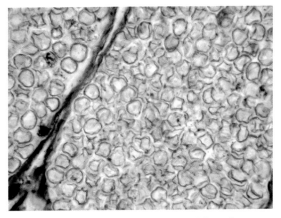

Fig. 299.   *Parataxodium* sp. pollen sacs containing pollen.

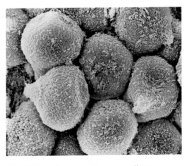

Fig. 300.   *Parataxodium* sp. pollen.

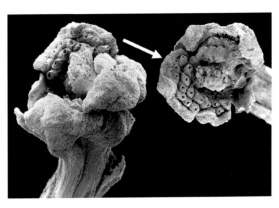

Fig. 301.   *Parataxodium* sp. immature seed cone and highlighting bract with ovules. Note two rows of ovules per bract.

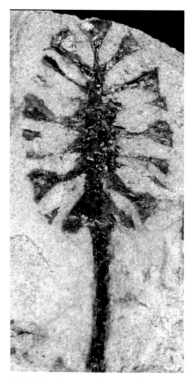

Fig. 302. *Parataxodium* sp. carbon trace seed cone.

Fig. 303. *Parataxodium* sp. calcite seed cone fractured tangentially on bracts.

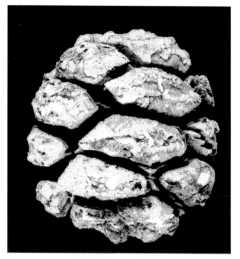

Fig. 304. *Parataxodium* sp. silicified seed cone.

Fig. 305. *Parataxodium* sp. silicified seed cone with seeds.

Fig. 306. *Parataxodium* sp. prepared block with degraded silicified seed cone.

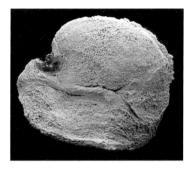

Fig. 307. *Parataxodium* sp. seed.

Fig. 308. *Parataxodium* sp. insect gall on leaf.

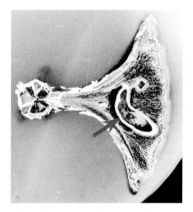

Fig. 309. *Parataxodium* sp. seed cone bract section showing insect boring (arrow).

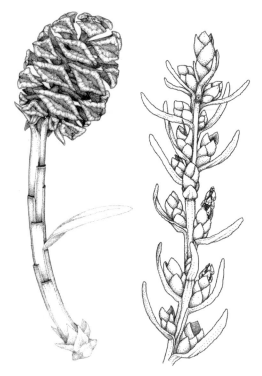
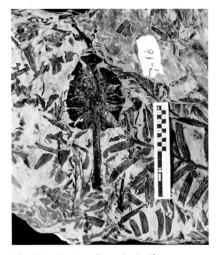

Fig. 310. Drawing of *Parataxodium* sp. seed cone and pollen cone branch from Horseshoe Canyon Formation.

Fig. 311. *Parataxodium wigginsii* from Alaska.

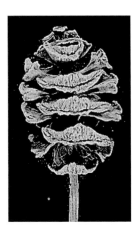
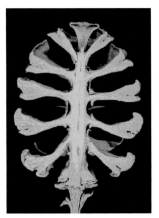

Fig. 312. *Metasequoia glyptostroboides* seed cone.

Fig. 313. *Metasequoia glyptostroboides* seed cone cross-section.

Fig. 314. *Metasequoia glyptostroboides* foliage, ventral view.

FOSSIL PLANTS

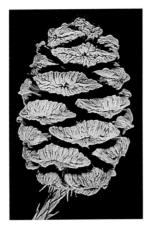

Fig. 315. *Sequoiadendron giganteum* seed cone.

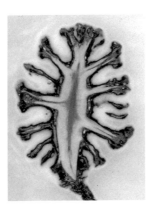

Fig. 316. *Sequoiadendron giganteum* seed cone cross-section.

Fig. 317. *Sequoiadendron giganteum* foliage.

Fig. 318. *Sequoia sempervirens* seed cone.

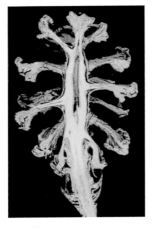

Fig. 319. *Sequoia sempervirens* seed cone cross-section.

Fig. 320. *Sequoia sempervirens* foliage, ventral view.

Fig. 321.   *Athrotaxis cupressoides* seed cone.

Fig. 322.   *Athrotaxis selaginoides* seed cone.

The Enigma of Metasequoia and Its Bearing on Phyllotactic Evolution in the Taxodiaceae/ Cupressaceae

Many mathematical papers and books have been written on phyllotaxy. Unfortunately these are beyond the comprehension of the amateur seeking general information. What follows is a simplified new interpretation based on the fossil *Parataxodium* and living *Metasequoia*.

Presently *Metasequoia* (Figs. 323, 324) has not been proven to exist in the Cretaceous based on the attributes of the living genus. Although this may seem unusual, it is directly relatable to the occurrence of *Parataxodium*. Both of them are identical in deciduous habit, but one is basically helical in disposition of organs and the other pseudo-opposite decussate. Very few traits set these two genera apart such as: the number and rows of ovules per bract in the seed cone; up to eight in a single row in *Metasequoia* and up to thirteen in two rows in *Parataxodium*; certain cuticular morphology; wavy anticlinal walls in the subsidiary cells of the leaf in *Metasequoia* and straight anticlinal walls in the subsidiary cells of the leaf in *Parataxodium*; and, of course, phyllotaxy; opposite decussate in *Metasequoia* and helical in *Parataxodium* (compare Figs. 324, 325).

If one considers them to be related, which they apparently seem to be, how do we go from a helical phyllotaxy (spiral of leaves and cone scales) to that of visually opposite decussate? How do we explain the aberrant vasculature noted by many previous authors (Morley 1948; Sterling 1949; Greguss 1956; Schwarz and Weide 1962; Bocher

Fig. 323. *Metasequoia glyptostroboides* foliage.

Fig. 324. *Metasequoia glyptostroboides* close-up of foliage.

KEVIN R. AULENBACK

Fig. 325. *Parataxodium* sp. silicified foliage.

1964; Namboodiri and Beck 1968b; Harris 1976; Saiki 1991; Takaso and Tomlinson 1992) for *Metasequoia*?

In plants we know that a certain number of leaves in a growth helix, by vascular connection, are produced before the first leaf is repeated (i.e., in a plant that produces thirteen leaves the next or the fourteenth is connected to number one by a vascular strand). We know how many strands divide in a given growth series as well as which strands in each series divides (Jean 1994).

Plants grow through a series of cycles to produce mature foliage. In a plant with thirteen leaves five from the lower series of eight had to divide. We know which strands divide as the plant grows and whether they diverge to the left or right. This is called the Lestiboudoise-Bolle theory of induction (Jean 1994) (Fig. 326).

Also basic to all plants is Fibonacci's number or the golden mean of 137.5°. This angle is the basis for plant growth. It appears to allow leaves to be produced forever without overlap. This can be shown as a cross-section of a hypothetical plant. If we place the theory of induction into this cross-section we get Figure 327.

Unfortunately, many plants deviate from this form. A range from 135 to 140° (compare Figs. 334 and 335 to Fig. 336) in vascular divergence is found in the Taxodiaceae. An individual plant species picks only one angle for divergence. For *Sequoiadendron giganteum* and *Parataxodium* from Drumheller it is 135° with thirteen vascular strands in mature growth or a 135, 5/13 series (Figs. 328, 334, 335).

Both the 137.5° growth and the 135° angle of divergence occur simultaneously in the same plant throughout all series of growth and can be shown by overlapping the two series (Fig. 329). The 137.5°

FOSSIL PLANTS

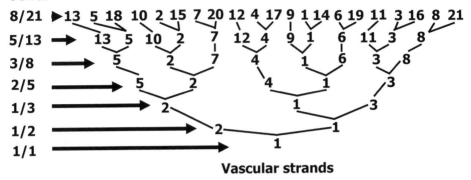

Fig. 326. Lestiboudoise-Bolle theory of induction.

angle is where the plant is most stable in growth and the divergence angle is where the leaf vasculature departs (Fig. 338). This causes the vascular strand to waver in the stalk. Waver of the vascular strands is recorded in *Sequoiadendron* and many other plants (Namboodiri and Beck 1968a; Kumazawa 1972).

How is torsion on the system controlled? Eventually the system would have one angle fall so far behind the other it would never catch up. It appears the next higher growth series first angle of divergence is a reset, so the whole system is spiralling ever so slowly upward, always catching up to itself. This allows for a repeat of any growth series to occur (Fig. 338).

We can now deal with a mature growth series of 5/13, 135° divergence (Figs. 330, 339), which occurs in both *Parataxodium* and *Sequoiadendron*. This cross-section can now be shown as a chart that mimics a cut-open cylinder of the branch. Any leaf vascular trace and divergence can now be plotted (Figs. 338, 339). Random cross-sections can show where any leaf vasculature would be at any given time in the cylinder. Note though in

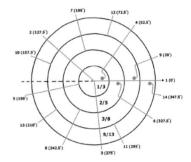

Fig. 327. 137.5° growth series ring up to 5/13.

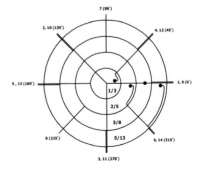

Fig. 328. 135° growth series ring up to 5/13.

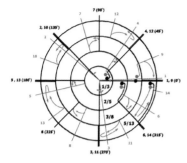

Fig. 329. Overlap of 137.5° and 135° growth series ring up to 5/13 with vascular waver direction arrows.

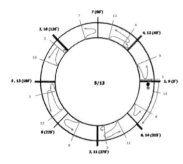

Fig. 330. Overlap of 137.5° and 135° in a 5/13 series with vascular waver direction arrows.

Figure 339 that there are still only eight departure angles yet thirteen vascular strands. In *Sequoiadendron*, the external phyllotaxy of 3/8 noted by others as problematic (Sterling 1945; Crafts 1943) with thirteen internal strands is solved with this model.

If this represents *Parataxodium*, how do we get the pseudo-opposite decussate nature of *Metasequoia* (Fig. 337)?

First, we must assume the basic premise that *Metasequoia* is a mature 135, 5/13 similar to *Parataxodium* and *Sequoiadendron* from which it originates. We now have assumed a spiral helix for the plant.

We could now try a multitude of ways to interpret *Metasequoia*'s phyllotaxy. Unfortunately, after many attempts, none worked except this; *Metasequoia* additionally now requires a shutting down of every alternate divergence or pole of 135° around the circle (Figs. 331, 332). Does this mess up the vasculature? Well, yes and no.

These alternate closures now force the departing traces to the open poles. The traces now push each other when trying to exit. This may seem chaotic for the vasculature in cross-section, but it is in fact very regular when drawn in a linear flat sheet (Fig. 340). These drawings explain all the anomalies seen by others (Morley 1948; Sterling 1949; Greguss 1956; Schwarz and Weide 1962; Bocher 1964; Namboodiri and Beck 1968b; Harris 1976; Saiki 1991; Takaso and Tomlinson 1992), who previously investigated the phyllotaxy in *Metasequoia*, from (false) double vascular stand departures, vascular strand fusion (push and shove), and vascular strand groupings.

These closures move and alternate after each full 5/13 expression with the result of eight sets of thirteen vascular strands of which the last four sets are mirror images of the first four (Fig. 333). Again one can plot any given leaf departure with regularity similar to *Parataxodium* or *Sequoiadendron*, except now there is a lot of push and shove prior to departure.

Metasequoia has in the past been interpreted as a type II and III phyllotaxy (Namboodiri and Beck 1968b) that is related to conifers with whorled or opposite/decussate phyllotaxy (the Cupressaceae). Upon closer inspection, *Metasequoia* still has a helical phyllotaxy similar to *Sequoiadendron*, but due to growth requirements now displays its foliage and reproductive structures in a pseudo-opposite/decussate fashion.

Metasequoia has many other specific traits that can be used to separate it from the rest of the Taxodiaceae. The erroneous use of deciduous traits, of which other Taxodiaceae (*Taxodium* and *Glyptostrobus*) share and which *Metasequoia* has been grouped with in the past, should not be used as singular proof of its existence in the fossil record.

The fossil *Parataxodium* in the Horseshoe Canyon is a deciduous *Sequoiadendron*. Its seed cones share many traits with *Sequoiadendron*; the only major difference is its deciduous nature.

Parataxodium is directly related to *Metasequoia* from which the latter evolved soon after the Cretaceous/Tertiary event. The reasons are unknown, but the evolution appears to have been rapid. Although the conifer retained its deciduous nature, its phyllotaxy has been altered as well as certain reproductive strategies.

The phyllotactic interpretations given, with variations in divergence, maturation and with or without alternate pole closure, can be further used to explain and unify phyllotaxy in all the remaining Taxodiaceae/Cupressaceae and Taxaceae (including false or pseudo-bijugacy; reinterpreted here as

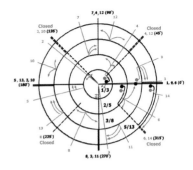

Fig. 331.   Overlap of 137.5° and 135° in a growth series ring up to 5/13 with alternate polar closures with vascular waver direction arrows (red), old departures (blue).

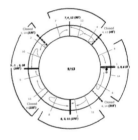

Fig. 332.   Same as Fig. 331 but mature 5/13 ring only showing leaf sets in pseudo-opposite decussate manner with vascular waver direction arrows (blue).

Foliar Expression Sets

Opposite

Leaf 1 (0°)   **Leaf 2 (180°)**

.../13, 10, 5, 2/ 10, 7, 2/ 12, 7, 4/ 12, 9, 4, 1/ 9, 6, 1   2, 5, 10, 13\ 5, 8, 13\ 3, 8, 11\ 3, 6, 11\ 1, 6, 9\...

Decussate

Leaf 3 (275°)   **Leaf 4 (90°)**

.../ 12, 7, 4/ 12, 9, 4, 1/ 9, 6, 1/ 11, 6, 3/ 11, 8, 3   4, 7, 12\ 2, 7, 10\ 2, 5, 10, 13\ 5, 8, 13\ 3, 8, 11\...

vascular series

.../---5----/-----4-----/----3----/----2----/-----------1------------\----2----\-------3--- --\----4---\----5----\...

Fig. 333.   Foliar expression sequence chart for *Metasequoia glyptostroboides.*

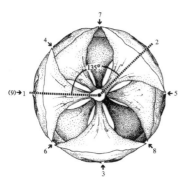

Fig. 334.   Drawing of *Parataxodium* sp. seed cone horizontal cross-section. 135° bract departures.

135°, 8/21 phyllotaxy) as well as a large segment of the Angiospermae. It can also be used to explain aberrant growth in the Angiospermae, such as observed in *Dipsacus silvestris* by Jean (1994), which suggests a 135°, 5/13 phyllotaxy alternate pole closure similar to *Metasequoia*, changing to a single open pole only.

Fig. 335.   *Sequoiadendron giganteum* seed cone horizontal cross-section. 135° bract departures.

Fig. 336.   *Cunninghamia lanceolata* seed cone horizontal cross-section. 140° bract departures.

Fig. 337.   *Metasequoia glyptostroboides* seed cone horizontal cross-section. Modified 135° bract departures.

FOSSIL PLANTS

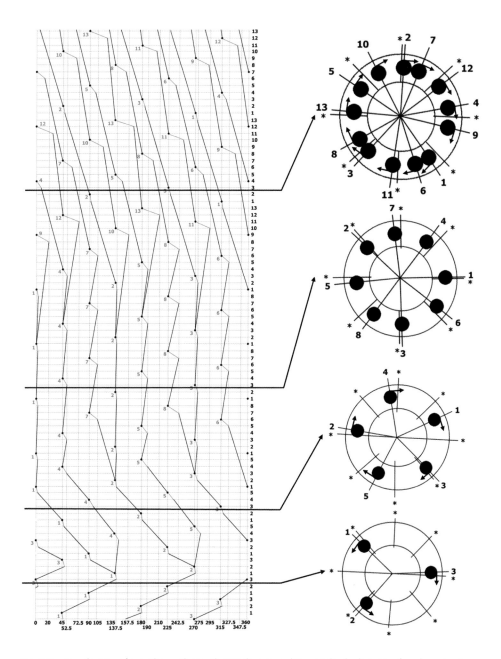

Fig. 338. Line drawing of growth in a plant with 135° departures (*Sequoiadendron*) starting from lowest series, 1/3, and ending at a 5/13. The right hand drawings represent vascular cuts through the stem at different series but same leaf departure; vasculature = black circles; arrows = direction vascular bundle is going at time of cut.

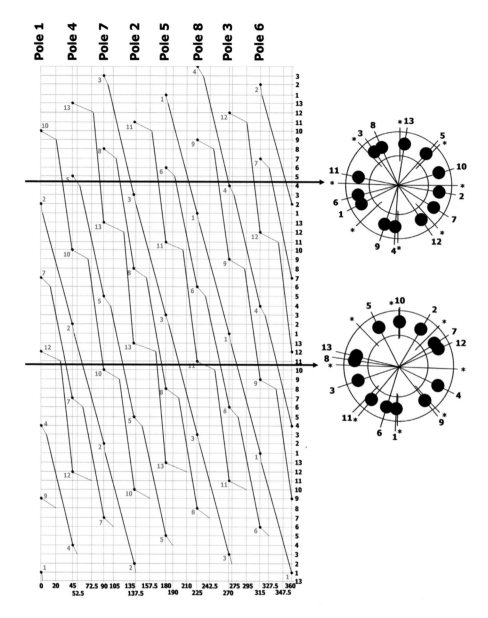

Fig. 339. Line drawing of a flattened cylinder of simplified vasculature in a 135°, 5/13 stem (*Sequoiadendron giganteum* and *Parataxodium* sp. from Drumheller). Two right-hand drawings represent random vascular cuts through the stem. Note the regularity of departures and the doubling up of vasculature over preceding departures in the flat cylinder. This gives *Sequoiadendron* its mistaken external count by previous authors of 3/8 phyllotaxy, yet in cross-sections it is actually a 5/13.

FOSSIL PLANTS

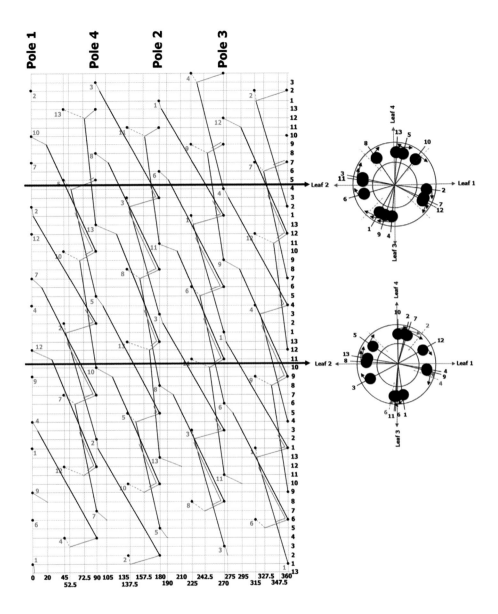

Fig. 340. Line drawing of a flattened cylinder of simplified vasculature in a modified 135°, 5/13 with alternate departure closures in a stem (*Metasequoia glyptostroboides*). Two right-hand drawings represent same cuts through the stem as Fig. 339; vasculature = black circles; arrows = direction vascular bundle is going at time of cut. Within flat chart green lines = pushing of two vascular strands; black lines = return of vasculature within cylinder; red lines = departure to leaf; red dotted line = normal departure for leaf; leaf departures = black dots. Note the regularity of the pattern in the flattened cylinder and the false chaotic appearance of the cross-cuts and apparent grouping of the vascular strands in juxtaposition to departures.

Fig. 341. *Taxodium distichum* foliage.

Fig. 342. *Taxodium distichum* close-up of foliage.

Fig. 343. *Taxodium wallisii* foliage.

Taxodioideae

CRYPTOMERIEAE

Taxodium (Swamp cypresses)

Over the last twenty to thirty years, it has been assumed that *Taxodium* (Figs. 341, 342) has existed in the area most commonly in the misidentified form of *Parataxodium*, which was discussed previously.

The first true and geologically oldest *Taxodium* was discovered in the Drumheller valley in 1998 and named *Taxodium wallisii* Aulenback and LePage, after the founder of the site, Mr. Kent Wallis. This is the earliest occurrence of *Taxodium* in the fossil record (Figs. 343–364).

This fossil *Taxodium* is well-represented in the Drumheller area. Silicified seed cones and pollen cones as well as calcite seed cones are quite common in the mudstones, although for the layman they may be difficult to identify. Branches with leaves are quite rare and properly identified wood is unknown. Although wood is identified from this species (Serbet 1997), this wood is thought to be from the extremely common *Parataxodium* species due to branch abscission base scars and associated leaves.

Leaves of the fossil *Taxodium* are decurrent, helically arranged, and of two types; scale-like leaves are restricted to pollen cone axes (Figs. 346–348) and vegetative leaves are linear falcate up to 11.5 mm long and 1 mm wide (Fig. 343). Stomata are randomly disposed on the adaxial face and randomly orientated to the leaf axis (Figs. 344, 345).

The fossil seed cones are helically arranged and held on branches bearing only seed cones

(Figs. 356, 364). Seed cones are short-stalked, round to oval in form up to 11 mm in diameter, and composed of up to fifteen deciduous, imbricate bract-scale complexes (Figs. 358, 359). The ovuliferous scales, which are prominently lobed, contain two seeds each (Figs. 357, 360, 361).

Dispersed seeds are wedge-shaped (Figs. 362, 363) and in many cases are still attached in pairs to the bract (Fig. 361).

Pollen cones were held in panicles similar to those of *Parataxodium*, but in *T. wallisii* the subtending leaves of the cones are helically disposed and scale-like, while the branch subtending leaf is linear falcate (Figs. 348, 364). The sporophyll contains five to nine microsporophylls in two rows (Figs. 349–353). Pollen is round with an apical papillae similar to *Parataxodium* (Fig. 354), which are both similar to dispersed pollen (Fig. 355).

The rare occurrence of fossil foliage would indicate that the tree may have been evergreen to semi-evergreen. This is similar to recent *T. mucronatum*, which is considered semi-deciduous, although the fossil shares more in common with *T. ascendens*, based on morphology. Both extant *T. ascendens* and *T. distichum* are considered fully deciduous.

Taxodium is closely related to *Glyptostrobus* and *Cryptomeria* of the Taxodiaceae (Figs. 365–376). Living *Taxodium* is represented by three species: *T. ascendens* (Pond Cypress), *T. distichum* (Swamp Cypress), and *T. mucronatum* (Mexican Swamp Cypress), all of which are monoecious and exist in moist to wet environments.

Fossil pollen:
Sequoiapollenites papillapollenites
Sequoiapollenites paleocenicus
Taxodiaceaepollenites hiatus

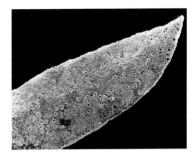

Fig. 344.   *Taxodium wallisii* linear leaf.

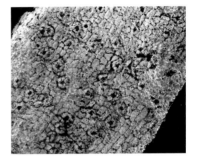

Fig. 345.   *Taxodium wallisii* close-up of Fig. 344 showing stomatal orientation and distribution.

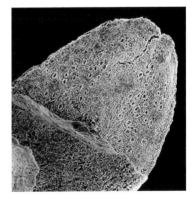

Fig. 346.   *Taxodium wallisii* pollen cone panicle scale leaf.

Fig. 347.   *Taxodium wallisii* close-up of
         scale leaf showing stomatal
         orientation and distribution.

Fig. 348.   *Taxodium wallisii* pollen
         cone panicle.

Fig. 349.   *Taxodium wallisii* partial pollen
         cone, apex removed.

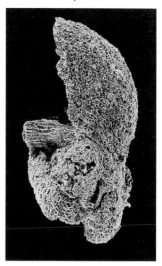

Fig. 350.   *Taxodium wallisii* single
         pollen cone sporophyll,
         lateral view.

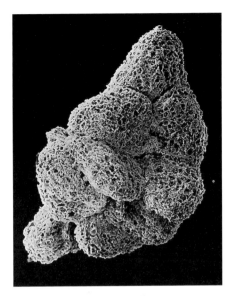

Fig. 351. *Taxodium wallisii* pollen cone sporophyll, basal view. Note 9 pollen sacs in two rows (compare to Fig. 298).

Fig. 352. *Taxodium wallisii* pollen cone medial section.

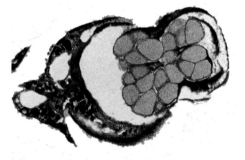

Fig. 353. *Taxodium wallisii* pollen cone tangential section.

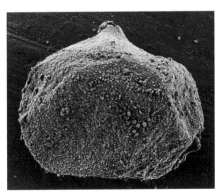

Fig. 354. *Taxodium wallisii* pollen.

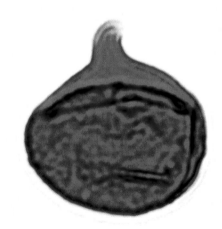

Fig. 355. *Sequoiapollenites paleocenicus* pollen.

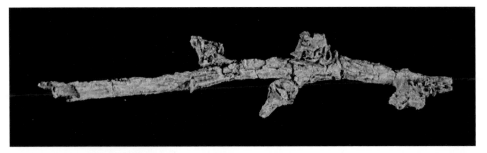

Fig. 356.   *Taxodium wallisii* seed cone branch bearing stalk remnants.

Fig. 357.   *Taxodium wallisii* immature seed cone, apical view.

Fig. 358.   *Taxodium wallisii* complete mature seed cone.

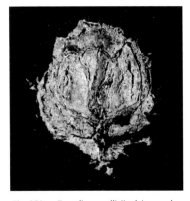

Fig. 359.   *Taxodium wallisii* calcite seed cone cross-section.

Fig. 360.   *Taxodium wallisii* seed cone bract without seeds, ventral view.

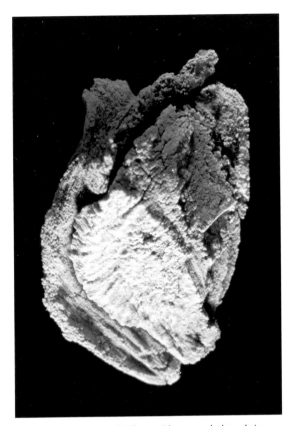

Fig. 361. *Taxodium wallisii* bract with two seeds, lateral view.

Fig. 362. *Taxodium wallisii* seed. Seed apex to the upper left.

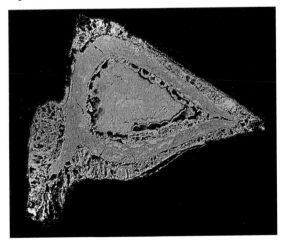

Fig. 363. *Taxodium wallisii* seed cross-section.

Fig. 364. Drawing of *Taxodium wallisii* pollen cones (left) and seed cones (right).

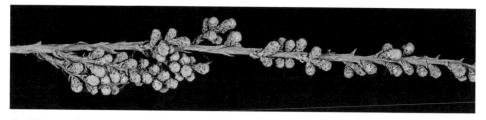

Fig. 365. *Taxodium distichum* pollen cone panicle.

FOSSIL PLANTS

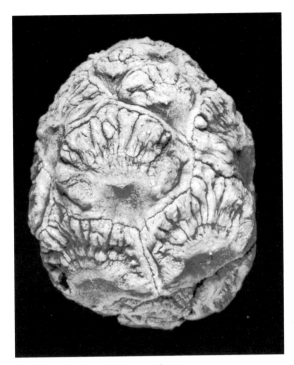

Fig. 366. *Taxodium distichum* seed cone.

Fig. 367. *Taxodium ascendens* seed cone.

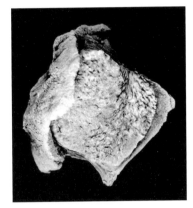

Fig. 369. *Taxodium distichum* bract with seeds, lateral view.

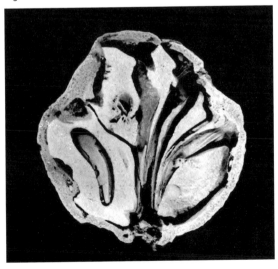

Fig. 368. *Taxodium distichum* seed cone cross-section.

Fig. 370. *Taxodium distichum* seed cross-section.

Fig. 371. *Cryptomeria japonica* seed cone.

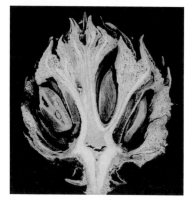

Fig. 372. *Cryptomeria japonica* seed cone cross-section.

Fig. 373. *Cryptomeria japonica* bract with four seeds.

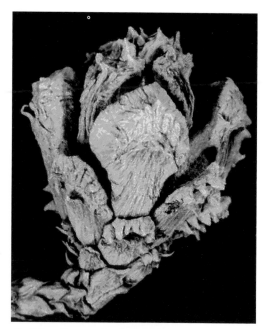

Fig. 374. *Glyptostrobus lineatus* seed cone.

Fig. 375. *Glyptostrobus lineatus* bract with two seeds.

Fig. 376. Phylogeny of seed cone development in selected Taxodiaceae, Cryptomeriae. Within the *Taxodium, Cryptomeria, Glyptostrobus* clade can be seen an evolutionary progression in the seed cones. *Taxodium* appears fairly straightforward. *Cryptomeria* today has actually two naturally occurring forms, *C. japonica* with from 3–5 seeds per bract from Japan and the lesser known *C. japonica* var. *sinensis* from Yunnan China with two seeds per bract. In the fossil record, it appears, based on seed attachment morphology, that two forms of *Glyptostrobus* can be recognized. Due to the complexity of the morphology and forms found in the clade, it is hypothesized that a prior extinction has occurred after the early Paleocene with subsequent new forms arising later. The clade gives us an unprecedented view of evolution in progress and a view of sporadic production of similar forms over time. Is *Cryptomeria* still evolving *Glyptostrobus*-like forms today? All cone bracts are drawn to scale.

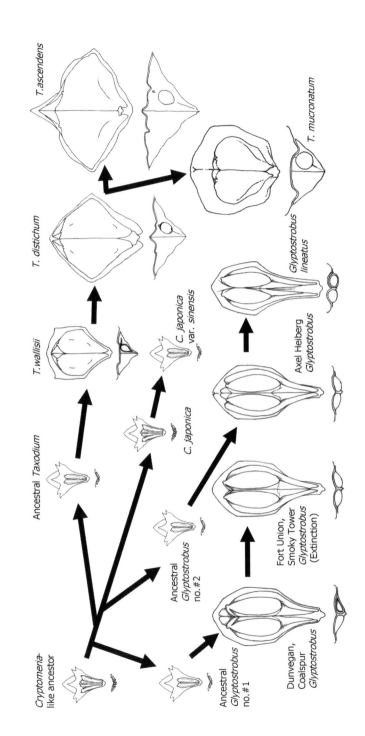

T. ascendens

T. distichum

T. mucronatum

T. wallisii

Ancestral *Taxodium*

C. japonica var. sinensis

C. japonica

Glyptostrobus lineatus

Axel Heiberg *Glyptostrobus*

Cryptomeria-like ancestor

Ancestral *Glyptostrobus* no.#2

Ancestral *Glyptostrobus* no.#1

Fort Union, Smoky Tower *Glyptostrobus* (Extinction)

Dunvegan, Coalspur *Glyptostrobus*

Fig. 377. *Cunninghamia lanceolata* seed cone.

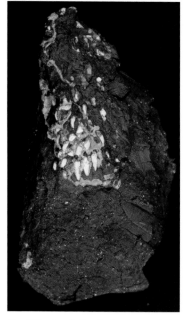

Fig. 378. Fossil *Cunninghamia* sp. siderite (ironstone) encased branch specimen prior to acid etching.

Taxodioideae

CUNNINGHAMIEAE

Cunninghamia (China fir)

If any living genus of the Taxodiaceae is more deserving of the name "living fossil," *Cunninghamia* is it. *Cunninghamia*-like conifers first occur in the Triassic in the foliage form genus *Elatides* (Stewart 1987). By the Cretaceous, forms such as *Cunninghamites* are readily comparable to the living genus. In the Horseshoe Canyon Formation, fossils are assignable to the living genus.

Although not as common as *Parataxodium*, *Cunninghamia* (Fig. 377) is well-represented by silicified and calcified seed cones and associated branches and are named a *Cunninghamia* species (Serbet 1997). Isolated seed cones were previously named *Pityostrobus* (*Cunninghamiostrobus*?) sp. (Bell 1949).

Branches are easily identified in the field as internal casts by their characteristic triangular leaf bases and well-defined stomatal bands. Calcite and silicified branch remains are common with stout helically arranged leaves up to 2.3 cm long with incurved tips (Figs. 378–380). Stomata are confined to the abaxial surface of the leaf in two distinct longitudinal rows that extend from the base to almost the apex. The leaf margins have minute teeth.

Seed cones are easily identified by their overlapping scales and characteristic size, up to 3 cm long and globose shape (Figs. 382–384). Bract scale complexes are helically arranged and contain a constricted stalk. On the inside of the bract/scale complex near the base is the ovuliferous scale whose margin is finely ornamented and show the position for the three attached inverted ovules. The outer edge of the bract is highly ornate with spine extensions (Figs. 381, 384). These spines are even viewable in calcite specimens encased in mudstone when fractured on the scales (Fig. 382).

There are only two living species of *Cunninghamia* presently recognized, *C. konishii* and *C. lanceolata*. *Cunninghamia* is a large monoecious, evergreen conifer with stiff decurrent non-petioled leaves with distinct dorsal stomatal bands. Pollen cones are borne terminally in groups and seed cones are terminal but solitary.

Known in Chinese as Sha-Shu, *Cunninghamia* is found throughout warm temperate China from sea level to two thousand metres. It occurs sporadically in western Hubei, western Szechuan, Hunan, Fujian, Yunnan, and Kiangsi provinces, as well as Taiwan. *Cunninghamia* is drought-tender but frost-resistant.

Fossil pollen:
Sequoiapollenites papillapollenites
Sequoiapollenites paleocenicus
Taxodiaceaepollenites hiatus

Fig. 379.   Fossil *Cunninghamia* sp. acid-etched specimen from Fig. 378.

Fig. 380.   Fossil *Cunninghamia* sp. naturally weathered branch cross-section.

Fig. 381. Mature silicified fossil bract (on left) compared to extant bract (on right).

Fig. 382. Fossil *Cunninghamia* sp. calcite replaced seed cone, weathered cross-section.

Fig. 383. Fossil *Cunninghamia* sp. calcite replaced seed cone roll out.

Fig. 384. Fossil *Cunninghamia* sp. multiple views of a silicified seed cone.

FOSSIL PLANTS

CUNNINGHAMIEAE

Taiwania-like

Taiwania-like three-dimensional needles are rare
and have been found at only three sites to date.
Two of the sites are in silicified mudstone (Figs.
388–396, 398–401) and the other in sandstone
(Figs. 387, 397). The leaf form is of two types, linear
falcate and cupressoid.

Linear leaves are decurrent, falcate, up to 1.2
cm long, and taper strongly to an obtuse pointed
apex. Stomata are confined to the facial (adaxial)
surface in two distinct bands (Figs. 388, 389). Stomata are transverse to the longitudinal axis of the
leaf and are alternate to opposite in one to four
rows per band (Figs. 389, 390).

Scale leaves show a gradation up to linear
forms and are also decurrent with the axis. These
leaves show similar stomatal rows but the apical
stomata are in oblique positions due to the curva-
ture of the leaf (Figs. 391, 392).

All leaves contain a proximal median resin
canal internally (Figs. 393, 394).

If the leaves were placed into a fossil taxon, it
would be in the form genus *Elatides* as they are
very similar to *Elatides williamsonii* from the
Yorkshire Jurassic of England (Harris 1979). This
genus of four species is presently only known from
the middle Jurassic to the Lower Cretaceous (Miller and Lapasha 1984).

Elatides was placed in the Taxodiaceae based
on similarities to modern genera in the family and
identified as containing both *Cunninghamia* and

Fig. 385.   *Taiwania cryptomerioides* foliage.

Fig. 386.   *Taiwania cryptomerioides* foliage close-up exhibiting dorsal and ventral stomatal bands.

Fig. 387.   Fossil foliage, carbon trace in sandstone.

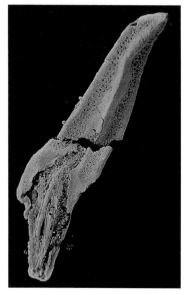

Fig. 388.   Facial view of silicified, flattened falcate needle.

Taiwania-like foliage forms (Miller and Lapasha 1984).

Elatides is presently known from the Lower Blairmore Group and Luscar Formation (Lower Cretaceous) in Alberta (Wan 1996). The fossil specimens from Drumheller would seem to indicate that *Elatides williamsonii* was at least still in existence in isolated areas in the Upper Cretaceous until the middle of the Horseshoe Canyon Formation, coal zone eight (Late Campanian).

In the living Taxodiaceae, the fossil leaf attachment to each other is similar to *Cryptomeria* or *Taiwania*, but the stomatal pattern and disposition is enigmatic. *Taiwania* differs in the stomatal band disposition and distribution (Figs. 385, 386). In *Taiwania* stomata are aligned parallel to the longitudinal axis of the leaf, and, although they are most prominent on the abaxial surface, they are also found on the adaxial surface.

At two silicified ironstone sites seed-cone bracts have been extracted, and in thin-sections of ironstones seed cones have been found. The seed-cone bract morphology appears most similar to *Taiwania* (Figs. 398–401).

The fossil seed cones consist of spirally inserted, persistent, imbricate scales. The flattened seed scale width expands up to 7 mm as it reaches the outer surface and then tapers quickly to a sharp upward point (Fig. 399). Each scale appears to have contained two inverted seeds based on seed scars (Figs. 399, 400).

At a sandstone site, carbon-trace seed cones were found attached to branches containing subtending leaves identical to those from silicified sites mentioned above, which confirms the connection of the silicified leaves and seed cone cross-sections. Whole seed cones appear to have been quite small, up to 1 cm long (Fig. 397).

The seed cones are similar to the Cretaceous fossil *Parataiwania nihongii* Nishida, Nishida, and Ohsawa from Japan, or *Athrotaxites berryi* Bell from North America. Unlike *P. nihongii*, which contains up to four seeds per bract, this fossil contains only two. *P. nihongii* at present does not have associated vegetative material for comparison (Nishida et al. 1991a).

The Horseshoe Canyon fossil, which contains both linear falcate and scale-like foliage, is unlike *A. berryi*, which contains only scale-like foliage and has longitudinally orientated stomata (Miller and Lapasha 1984). Seed cones of *A. berryi* contain a single seed per bract and smaller seed cone scales (2.5 mm wide) (Miller and Lapasha 1984).

It is speculative at this time to place these fossil remains in the extant genus *Taiwania* due to the stomatal morphology, but the seed cones appear to be *Taiwania*-like. The placement is therefore tentative until further comparative studies on the relationship and morphology of *Elatides* and *Taiwania* is undertaken.

Taiwania cryptomerioides (Figs. 385, 386) is a large tree in its native habitat growing up to fifty metres tall. At one time, it was thought to be restricted to the island of Taiwan, but it is now known to exist at a variety of isolated sites throughout central and southern China.

This fossil has not been formally described.

Fossil pollen:
Sequoiapollenites papillapollenites
Sequoiapollenites paleocenicus
Taxodiaceaepollenites hiatus

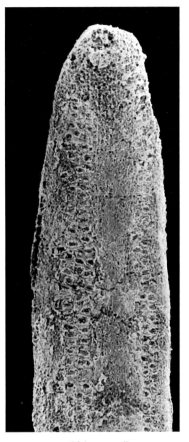

Fig. 389.   Fossil falcate needle apex.

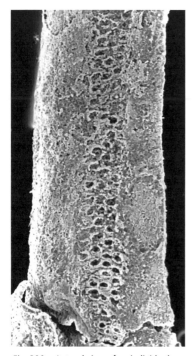

Fig. 390.   Lateral view of an individual
            stomatal band.

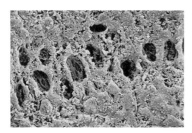

Fig. 392.   Cupressoid stomata from Fig.
            391.

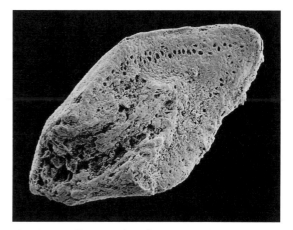

Fig. 391.   Fossil cupressoid needle.

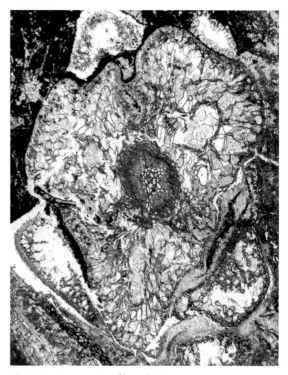

Fig. 393.   Cross-section of branch in ironstone.

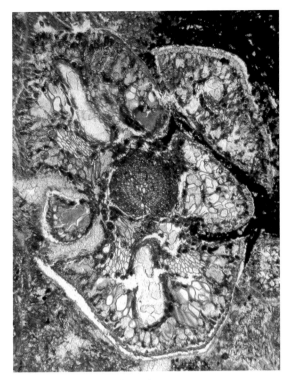

Fig. 394.   Cross-section of branch in ironstone.

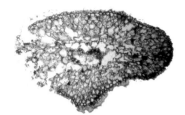

Fig. 395.   Cross-section of linear needle apex.

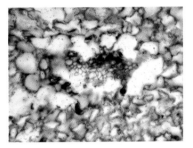

Fig. 396.   Vascular strand cross-section from Fig. 395.

Fig. 397.   Seed cones and foliage.

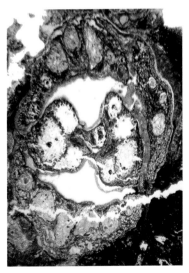

Fig. 398.   Cross-section of seed cone.

Fig. 399. Fossil seed cone bract with mature seed scar (left) and abortive seed scar (right) at arrows.

Fig. 400. Fossil seed cone bract bearing seed attachment scars (arrows).

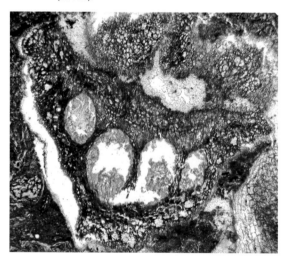

Fig. 401. Seed cone bract cross-section.

Taxodioideae

Athrotaxis-like

This Taxodiaceous genus has been found at a single site in the Horseshoe Canyon. The fossil is represented by helically disposed cupressoid foliage (Figs. 404, 407, 408) that was most prominent around the base of a silicified tree stump excavated near the town of East Coulee (Figs. 405, 406). Although the foliage form is common to many conifer genera, including the Podocarpaceae, this leaf form agrees in aspects to previously described fossils as *Athrotaxites* Unger, which is considered to belong to the Taxodiaceae (Miller and Lapasha 1984). The leaves have not been scanned for stomata at present and reproductive material appears absent. It presently differs from the previous *Elatides* or *Taiwania*-like material in that linear leaves are absent.

The tree stump that this foliage surrounds has been cored and sectioned. The cross-section shows distinct growth rings of extremely variable width. Latewood occupies from 1/6 to 1/3 of the ring and resin canals are absent (Figs. 409, 410).

The radial section shows vertical tracheids with pits in one to three rows, the most common being two. Pits are opposite and touching. Pit apertures are round to elliptical. Crassulae are common between vertical pit rows (Figs. 411, 412). Rays consist of only parenchyma. Ray cross fields contain from one to three pits in a single row (Fig. 413). The pits are bordered and apertures are slit-like, elliptical, and oblique in orientation (cupressoid). Marginal cells are present and contain

Fig. 402.   *Athrotaxis cupressoides* foliage.

Fig. 403.   *Athrotaxis cupressoides* foliage close-up.

Fig. 404. Fossil *Athrotaxis*-like foliage.

from two to seven pits in one to two rows per cross field (Figs. 413, 414). Pits are similar in size to normal ray cross fields, bordered and containing an oblique elliptical slit-like aperture (cupressoid).

The tangential section shows vertical tracheids with scattered simple bordered pits (Figs. 415, 416). The pits have circular to slit-like oblique apertures. Rays are uniseriate, rarely biseriate and from one to eleven cells high with two to four being most common. Vertical wood parenchyma appears common and is encountered more towards the latewood. Parenchyma has smooth end walls and commonly contains dark contents (resin) (Fig. 415).

The wood is Taxodiaceae and shows some similarity to *Athrotaxis* or possibly *Taiwania*. Due to its association of the stump with the foliage, it is considered to be from an *Athrotaxites*-like species. Hopefully future investigations of the foliage cuticular characters will confirm this identity.

In the living genus *Athrotaxis* (Figs. 321, 322, 402, 403), three species are found, all of which are endemic to the island of Tasmania. These are *A. cupressoides* (Pencil Pine), *A. laxifolia* (Summit Cedar), and *A. selaginoides*. All are monoecious evergreens capable of vegetative reproduction.

Athrotaxis cupressoides is 6–15 m tall and prefers partial to full shade with a rich soil (grass or sphagnum covered). It is frost-resistant but drought-tender and restricted to montane forests of Tasmania.

Athrotaxis laxifolia is up to 9 m tall and also prefers partial to full shade in rich soils. It was previously thought to be a hybrid between *A. cupressoides* and *A. selaginoides*, but this has been disproved via genetic studies. It is restricted to growing ranges between that of *A. cupressoides* and *A. selaginoides*.

Athrotaxis selaginoides grows up to 35 m tall and is found on the valley floors of central highlands.

Fossil pollen:
Sequoiapollenites papillapollenites
Sequoiapollenites sp.
Taxodiaceaepollenites hiatus

Along with the fossil stump, a wealth of ecological information was obtained. The site interpretation follows.

Fig. 405. Preliminary dig. D. Braman for scale.

Fig. 406. Stump after extraction. Technician A. Orosz for scale.

Fig. 407. Foliage pediment from around base of stump.

"Three-Tonne Toby": A Silicified Tree Stump Microenvironment Flora

In 1998 a fossil tree stump was placed as a showpiece in the Conservatory of the Royal Tyrrell Museum (Figs. 405, 406). The stump was removed from the lower Horseshoe Canyon Formation coal seam #1 just north of the town of East Coulee, Alberta, where it had originally grown during the Cretaceous approximately 72 million years ago. The stump was rooted into the lowermost quarter of the coal.

When collected, the specimen weighed 2,903 kg or 2.903 tonne but in life the stump itself would have weighed much less. Due to its massive weight, it was appropriately nicknamed "Three-Tonne Toby." A core sample and ring counts revealed that the tree had lived a minimum of 359+ years prior to death and subsequent burial.

In the area surrounding the tree were rhizomes in growth position of an *Osmunda*-like fern (see Fern section; *Osmunda*) as well as two fossil fern foliage types, one common and one rare. Sections of the rhizome show overall morphology similar to those of *Osmunda* found in the formation, although the preservation is coal and calcium phosphate instead of silica. Externally the rhizomes are long, rope-like and bifurcate (Fig. 417). There are no indications of large root masses.

The more common foliage appears to be *Osmunda*-like in form (Fig. 420). The much rarer fern pinnae sections have not been identified (Fig. 421). The remainder of the coal consists of amorphous plant parts.

The preservation of only a few recognizable plants amongst large amounts of amorphous

FOSSIL PLANTS

coaly debris is not unusual. In recent coal-forming swamps such as the *Taxodium* swamps of Louisiana or Florida, seasonal flooding constantly changes the landscape. In these low-lying flood plains, plant species may change radically with a lateral or vertical change of a few centimetres due to soil composition and water availability (Conner and Day 1976).

Soils in these areas are usually thin and plants rely on periodic seasonal flooding to replenish nutrients. Trees in these marginal areas are subjected to having their roots continually or sporadically underwater, but they support their own microflora about their trunks and roots by continually trapping detritus. Open areas between trees are usually barren and form an accumulation of transported, degraded, amorphous plant parts. This was very similar to that seen in the coal deposition that surrounded Three-Tonne Toby.

Cross-sections of the coals show the presence of alternating layers of fusain (Figs. 418, 419). This fusain represents burnt plant remains from periodic or cyclic forest fires. The fusain was most likely transported into the coal swamp from the surrounding area by seasonal flooding as many pieces show rounded edges indicative of tumbling. This appears to have been a common occurrence in the area just as cyclic or periodic fires are a natural part of forests today.

During the mining era in the Drumheller valley, stumps called "Niger Heads" similar to the fossil were frequently found in the mines. These slowed mining and damaged machinery. Stumps were tunnelled around or removed by hand to prevent sparks from metal striking the silica that could ignite the coal dust and cause mine explosions.

Fig. 408. Weathered foliage from pediment.

Fig. 409. Fossil wood cross-section from core.

Fig. 410. Fossil wood close-up of cross-section.

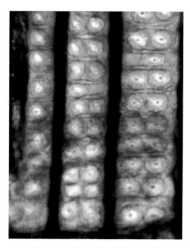

Fig. 411. Fossil wood radial view of tracheids.

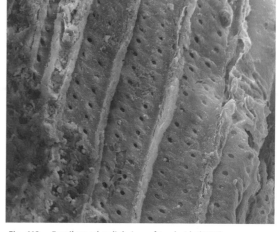

Fig. 412. Fossil wood radial view of tracheids (SEM).

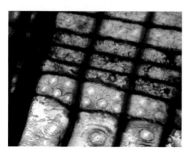

Fig. 413. Fossil wood radial view of cross-fields.

Fig. 414. Fossil wood radial view of cross-fields (SEM).

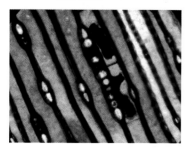

Fig. 415. Fossil wood tangential view of rays showing cells with resin contents.

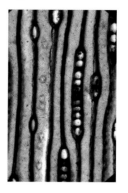

Fig. 416. Fossil wood tangential view of rays.

FOSSIL PLANTS

Fig. 417.   Fern rhizomes freshly exposed.

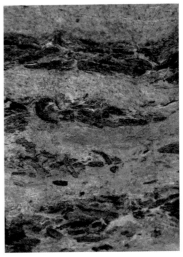

Fig. 418.   Lateral view of layered fusinites in pediment.

Fig. 419.   Fusinites, top view.

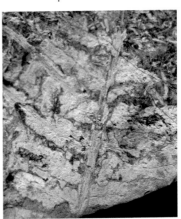

Fig. 420.   Common fossil fern pinnae.

Fig. 421.   Rare fossil fern pinnae.

CUPRESSACEAE (CYPRESS)

The living Cupressaceae are divided into the following groups:

Subfamily Callitroideae
  Tribe Callitreae
  Tribe Actinostobeae
  Tribe Libocedreae
  Tribe Tetraclineae

Subfamily Cupressoideae
  Tribe Cupresseae
  Tribe Thujopsideae
  Tribe Junipereae

The Cupressaceae is an ancient family of conifers that first occur in the Cenomanian. Other plant families from the Cretaceous bore cupressaceous foliage also, such as the extinct Cheirolepidiaceae, which contain such genera as *Frenelopsis* and *Brachyphyllum*. This family is, however, not related to the Cupressaceae.

Some Cretaceous fossil Cupressaceae are even placed in extant genera, although these can be seen as misidentifications. For example, *Thuja smileya* LePage, from the Turonian of Alaska, is easily re-interpreted as a pollen cone branch of *Parataxodium*.

The putative seed cones of *T. smileya* show no indication of scale-like subtending leaves and are in sediments containing many individual leaves of a *Parataxodium*-like plant which surround them. There is no indication of scale-like foliage on the subtending branch, in the block or in the sediments (LePage 2003). The "seed cones" present are held in disposition identical to pollen cones identified as *Parataxodium* from in the Horseshoe

Canyon. Misidentification such as this has previously occurred in the identification of *Juniperites gracilis* (Heer) Seward and Conway from in the Horseshoe Canyon Formation (Bell 1949), where upon viewing, the original figured specimen (Bell 1949, pl. IV, fig. 5) is presently identified as pollen cone branch of *Parataxodium*.

It is presently speculated that extant genera representing the Cupressaceae do not exist in the Cretaceous fossil record worldwide. Any claim of genus longevity in this ancient yet presently highly evolved group should be viewed as suspect.

Mesocyparis

Only one genus of Cupressaceae has been identified from the Horseshoe Canyon Formation, *Mesocyparis*.

This genus represents an extinct line of the Cupressaceae and has no living counterpart. It was abundant during the Upper Cretaceous in the fossil form *Mesocyparis umbonata* McIver and Aulenback and can be found at many different localities in the Horseshoe Canyon.

M. umbonata is represented in the fossil record by all plant parts except roots. These plant parts are preserved as silica, calcite, and carbon trace (Figs. 422–436, 438–447, 449–452) and appear to be very common in the area.

Leafy sprigs are fern-like in appearance (Figs. 422–427) and are most often identified as such due to the minute bifacially flattened leaves. Leaves are small up to 4 mm long and held opposite decussate on a dorso-ventrally compressed branch.

Pollen cones are small (Figs. 432–434) up to 2.1 mm long and found restricted to pollen cone branches. Cones are terminal but proximally positioned on the branch, which contains leafy foliage

Fig. 422. *Mesocyparis umbonata* carbon trace foliage.

Fig. 423. *Mesocyparis umbonata* silicified foliage.

Fig. 424. *Mesocyparis umbonata* silicified foliage.

distally (Fig. 448). Pollen cone sporophylls contain two microsporophylls each (Fig. 435). Pollen is round and lacking an apical papillae (Fig. 436) similar to *Taxodiaceaepollenites hiatus* (Fig. 437).

Seed cones are small (Figs. 438–445) up to 5 mm wide and also restricted to seed cone branches. Seed cones are borne opposite decussate on branch shafts that terminate in a leafy branch (Fig. 448). Seeds are small with two inflated lateral wings (Fig. 446).

Wood samples of *M. umbonata* have also been prepared (Figs. 449–452).

Vegetative plant remains in the formation were referred to *Thuites interruptus* Newberry (Bell 1949), but, unfortunately, the taxa was a catchall for fossil cupressoid leaves from various formations, which varied greatly in branching forms and leaf sizes. The fossil from the Horseshoe Canyon Formation did not match the type and was subsequently named *Mesocyparis umbonata*.

Mesocyparis is also represented in the early Paleocene of Saskatchewan by *Mesocyparis borealis* McIver and Basinger (1987), which is identical in foliage and reproductive organ disposition but much larger in size (this text). *Mesocyparis beringiana* (Golovneva) McIver and Aulenback, from the Cretaceous of eastern Russia, is identical in gross foliage morphology and seed cone disposition to *Mesocyparis umbonata* but contains differences in seed cone structure based on the original paper (Golovneva 1988).

Another new species that appears similar to both *Mesocyparis* and *Chamaecyparis* is *Mesocyparis rosanovii* (Kodrul et al. 2006). It bears seed cones identical in morphology and disposition to *Mesocyparis*, but its foliage disposition is more reminiscent of *Chamaecyparis*. *Chamaecyparis* is found in the Cretaceous of Vancouver Island

(McIver 1994). *Chamaecyparis corpulenta* seed cone morphology is very similar to *Mesocyparis*, although its seed cone disposition and foliage morphology are different.

The fossil genus *Mesocyparis*, as a whole, appears to be an intermediary between northern (*Chamaecyparis*) and southern (*Papuacedrus, Austrocedrus, Libocedrus*) hemispheric genera (Figs. 453–476). There are seven recent species of *Chamaecyparis*, one of *Austrocedrus*, five of *Libocedrus*, and three of *Papuacedrus*.

M. umbonata is thought to have been a low spreading shrub similar in form to the junipers that exist presently in the Drumheller valley or that of *Microbiota decussata* from southeastern Siberia.

Fossil pollen:
Same as for the *Taxodiaceae*

Fig. 425. *Mesocyparis umbonata* silicified foliage.

Fig. 426. *Mesocyparis umbonata* cross-section of foliage.

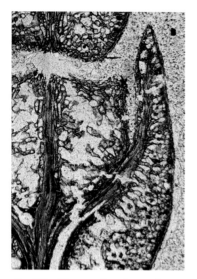

Fig. 427. *Mesocyparis umbonata* close-up of cross-section showing intact cells.

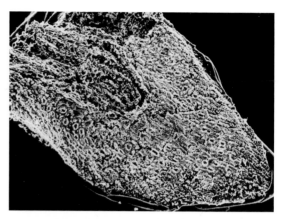

Fig. 428. *Mesocyparis umbonata* facial leaf.

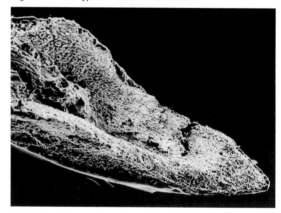

Fig. 429. *Mesocyparis umbonata* lateral leaf.

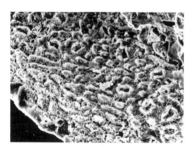

Fig. 430. *Mesocyparis umbonata* facial leaf stomatal close-up.

Fig. 431. *Mesocyparis umbonata* lateral leaf stomatal close-up.

Fig. 432. *Mesocyparis umbonata*
pollen cone.

Fig. 433. *Mesocyparis umbonata* pollen
cone close-up.

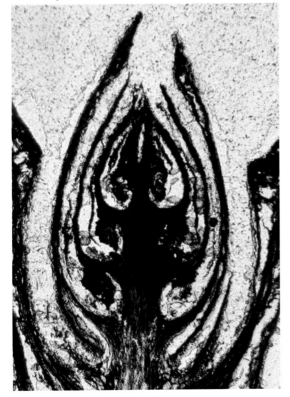

Fig. 434. *Mesocyparis umbonata* pollen cone cross-section.

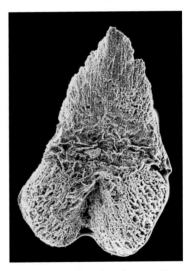

Fig. 435. *Mesocyparis umbonata* pollen
cone sporophyll.

Fig. 436.   *M. umbonata* pollen grain.

Fig. 437.   *Taxodiaceaepollenites hiatus* pollen.

Fig. 438.   *Mesocyparis umbonata* carbon trace seed cone branch containing seeds.

Fig. 439.   *Mesocyparis umbonata* carbon trace seed cone branch without seeds.

Fig. 440.   *Mesocyparis umbonata* silicified immature seed cone branch.

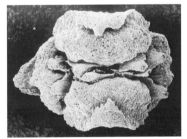

Fig. 442.   *Mesocyparis umbonata* slightly more mature seed cone with ovules, apical view.

Fig. 441.   *Mesocyparis umbonata* immature seed cone showing ovules, apical view.

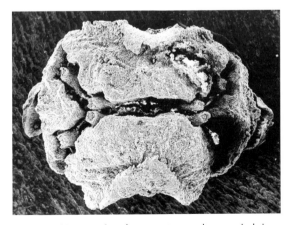

Fig. 444.   *Mesocyparis umbonata* mature seed cone, lateral view.

Fig. 443.   *Mesocyparis umbonata* mature seed cone, apical view.

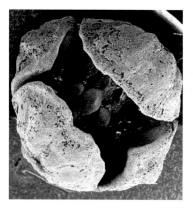

Fig. 445. *Mesocyparis umbonata* mature seed cone after seed shedding.

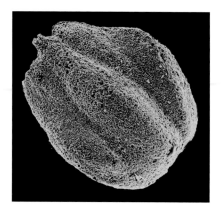

Fig. 446. *Mesocyparis umbonata* seed.

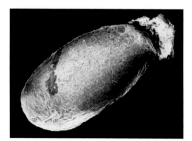

Fig. 447. *Mesocyparis umbonata* seed cuticle.

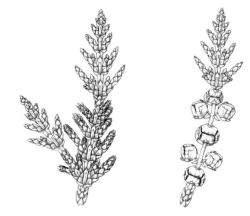

Fig. 448. Reconstruction of *Mesocyparis umbonata* pollen cone bearing branch (left) and seed cone bearing branch (right).

Fig. 449. *Mesocyparis umbonata* wood cross-section.

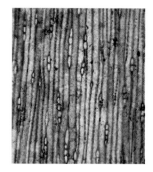

Fig. 450. *Mesocyparis umbonata* tangential view of rays.

Fig. 453. *Platycladus orientalis* foliage.

Fig. 451. *Mesocyparis umbonata* radial view of tracheids.

Fig. 454. *Platycladus orientalis* seed cone.

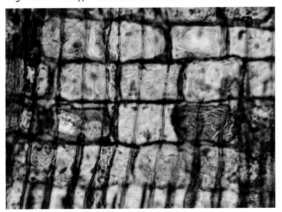

Fig. 452. *Mesocyparis umbonata* radial view of cross-fields.

Fig. 455. *Calocedrus decurrens* foliage.

Fig. 456. *Calocedrus decurrens* pollen cones.

Fig. 457. *Calocedrus decurrens* seed cone.

Fig. 458. *Chamaecyparis nootkatensis* seed cone.

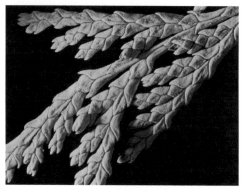

Fig. 459. *Chamaecyparis nootkatensis* foliage.

Fig. 460. *Chamaecyparis thyoides* seed cones.

Fig. 461. *Austrocedrus chilensis* pollen cones.

Fig. 462. *Austrocedrus chilensis* foliage.

Fig. 463. *Austrocedrus chilensis* seed cone.

Fig. 464. *Cupressus* sp. seed cone.

Fig. 465. *Cupressus* sp. foliage.

Fig. 466. *Thuja plicata* seed cone.

Fig. 467.   *Thuja plicata* foliage.

Fig. 468.   *Papuacedrus papuana* foliage.

Fig. 470.   *Papuacedrus papuana* seed cone.

Fig. 469.   *Papuacedrus papuana* foliage.

Fig. 471. *Papuacedrus papuana* branch bearing seed cones.

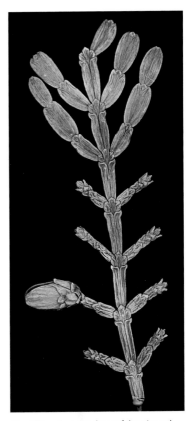

Fig. 472. *Papuacedrus arfakensis* seed cone bearing branch.

Fig. 473. *Libocedrus plumosa* seed cone.

Fig. 474. *Libocedrus plumosa* foliage.

Fig. 475. *Microbiota decussata* foliage.

Fig. 476. *Microbiota decussata* seed cones.

Fig. 477. *Podocarpus macrophylus*
immature seed cones.

PODOCARPACEAE (YELLOW WOODS)

Subfamily Pherosphaeroideae
Subfamily Phyllocladoideae
Subfamily Podocarpoideae

The Podocarpaceae is an ancient family that formed part of the Gondwanan flora (White 1990). They are presently found in Africa, South America, and Australasia.

The living Podocarpaceae, which contains approximately a hundred species, is divided into three subfamilies. These are: Pherophaeroideae (*Microstrobus*), Phyllocladoideae (*Phyllocladus*), and Podocarpoideae (*Podocarpus, Dacrydium, Dacrycarpus, Lagarostrobus, Microcachrys, Prumnopitys, Saxegothaea*, and *Acmopyle*).

Only pollen presently represents the Podocarpaceae in the formation (Fig. 479). With limited knowledge present for the pollen identifications and the diverse leaf forms presently in the Podocarpaceae (Figs. 477, 478, 480), it would be extremely speculative to suggest an affinity to any genera or form. This plant fossil appears to await discovery.

Although supporting evidence of Podocarpaceae in the Mesozoic northern hemisphere is sparse and not wholly reliable, Podocarpoid wood has been identified from the southern United States (Wheeler and Lehman 2005).

Fossil pollen:
Podocarpidites ellipticus
Podocarpidites multesimus

Fig. 478. *Dacrydium franklinii* foliage and immature seed cones.

Fig. 479. *Podocarpidites* sp. pollen.

Fig. 480. *Microcachyrs tetragona* seed cone and cupressoid branches with *Sellaginella* in the background.

Fig. 481.   *Torreya californica* foliage.

Fig. 482.   *Torreya californica* branch ventral view showing pollen cones.

Fig. 483.   *Torreya californica* close-up of leaf stomatal bands.

TAXOPSIDA

Taxales

TAXACEAE (YEWS)

The Taxaceae contains five genera: *Amentotaxus, Torreya, Austrotaxus, Pseudotaxus,* and *Taxus.* These genera occur in a variety of habitats from warm temperate to tropical.

Torreya

Isolated fossil needles and branches with attached needles of the Taxaceae have been recovered from the Horseshoe Canyon Formation. These specimens were considered morphologically and anatomically similar to the genus *Taxus* (Serbet 1997) (Fig. 487). Closer examination of silicified individual leaves as well as a sandstone sample with branch and attached leaves show these fossils to be that of a *Torreya* species.

The genus *Torreya* can be traced as far back as the Jurassic of England in the form of *Torreya gracilis* Florin, based on its highly diagnostic needles (Florin 1958; Harris 1979).

The fossil leaves from the Horseshoe Canyon are up to 2.5 cm long and 3 mm wide with a linear tapering apex and a constricted petiolar base (Fig. 484). The adaxial surface is smooth and devoid of stomata. Stomata are confined to the abaxial surface in two distinct narrow grooves (Fig. 485). Stomata are orientated perpendicular to the axis of the leaf (Fig. 486).

Pollen of fossil *Torreya* is included in the fossil Taxodiaceae form genera as it cannot be distinguished under the light microscope from the latter.

Seeds of *Torreya* have not yet been recognized from the formation. Living *Torreya* seeds are oval and contain a vascular pore on both sides towards the apex.

Presently in the genus *Torreya*, there are six living species in North America and China. *Torreya taxifolia* and *T. californica* (Figs. 481–483) inhabit North America, and *Torreya grandis*, *T. nucifera*, *T. jackii*, and *T. fargesii* inhabit China and Japan. *T. californica* inhabits riverbanks and valleys in California, while *T. taxifolia* is found only on the eastern banks of the Apalachicola River in Florida and sporadically in Georgia.

Torreya grandis grows from western Hubei, eastern Szechuan, Chekiang, and Fujian in mixed woodlands. *T. fargesii* grows in Szechwan and Hubei provinces in China in the mountains at approximately 1,400 m and *T. nucifera* grows from 1,000 to 4,000 m in Japan.

Various similarities in bark have been noted between *T. grandis* and *T. taxifolia* as well as between *T. nucifera* and *T. californica*.

Torreya, also called "stinking cedars" due to their aromatic smell when the leaves are crushed, are evergreen, dioecious (male and female on separate plants) trees. They are also drought tender.

This fossil *Torreya* has not been described and is deserving of a much more in-depth investigation.

Fossil pollen:
Sequoiapollenites papillapollenites
Sequoiapollenites paleocenicus
Taxodiaceaepollenites hiatus

Fig. 484.   Fossil carbon trace needle and recent needle of *Torreya californica.*

Fig. 485.   Silicified fossil needle (SEM).

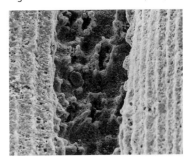

Fig. 486.   Fossil *Torreya* sp. close-up of stomatal groove.

Fig. 487.   *Taxus floridana* with pollen cones.

Fig. 488. *Torreyites tyrrellii* foliage block.

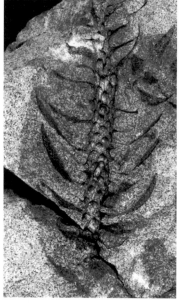

Fig. 489. *Torreyites tyrrellii* close-up of Fig. 488.

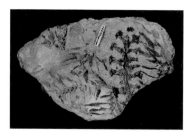

Fig. 490. *Torreyites tyrrellii* sample containing seed cone.

Incertae Sedis

Unknown Coniferous Plants and Pollen

Torreyites tyrrellii (DAWSON) BELL

Torreyites tyrrellii (Dawson) Bell, in the formation, is based on fossil branches preserved as carbon trace in sandstone (Figs. 488–491) and silicate and carbonate trace in ironstone. The fossil leaves are not petiolate like *Torreya* and it is obvious from first glance that it is not related.

The fossils attributed to this genus (Bell 1949) appear to be a mixture of *Cunninghamia* and other plants. Some, from various sites, are *Cunninghamia* and exhibit the prominent stomatal bands. Others, like the one illustrated, lack any defined stomatal bands. The leaves of these specimens are exceedingly long, up to 4 cm when compared to the other fossil *Cunninghamia* previously discussed.

If the seed cone on the sample specimen of *Torreyites tyrrellii* (Figs. 490, 491) belongs to the foliage, then this is indeed a new genus of conifer unrelated to *Cunninghamia* since the unidentified cone, although incomplete, is quite large (5.8+ cm long) and has peltate bracts (compare Figs. 491 and 492). It appears Taxodiaceous, but it is unknown what its relationship is at this time.

CONIFER OR ANGIOSPERM SEED?

This fossil seed type is commonly encountered in the formation as a silicified roll out from the sediments.

These large drupe-shaped seeds contain a pair of large apical vascular pores, one on each side. The seeds (Figs. 493, 494) are up to 9.5 mm long.

Originally thought to be a *Torreya* seed, based on the large paired apical vascular pores similar to those described as *Diploporus torreyoides* Manchester or possibly *Torreya clarnensis* Manchester from the Eocene of Oregon (Manchester 1994), this appears not to be the case. Seeds of these two genera vary in seed shape, size, and vascular pore placement when compared to the Horseshoe Canyon specimens.

The seed *Vesquia* from the Cretaceous of Belgium has its vascular pores situated in the lower portion of the seed, which is somewhat similar in form to *Diploporus* (Manchester 1994) and is unlike the seeds from the Horseshoe Canyon.

A half-seed internal cast, which appears cospecific with the Horseshoe Canyon seeds preserved as ironstone from the Oldman Formation of southern Alberta, contains an embryo that appears very unusual (Figs. 495, 496). The embryo appears centrally divided, reminiscent of an angiosperm dicot, but each cotyledon appears to divide again multiple times further up, which would indicate a conifer. The lateral strands from the pore areas are unusual and are not found in any extant or fossil conifer or angiosperm. The lateral strands appear embryo-like but the "radicle" appears firmly attached in the pore area.

Many plants can produce multiple embryos at early stages of development. In the fossil these laterals may represent abortive embryos but they appear very well-developed. The apical strands of these possible lateral abortive embryos may have attached to the central embryo as they also appear to divide higher up and follow closely the embryo.

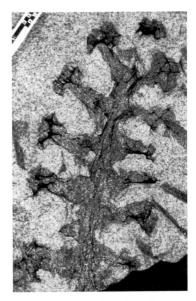

Fig. 491.   Close-up of seed cone from Fig. 490.

Fig. 492.   *Cunninghamia lanceolata* seed cone.

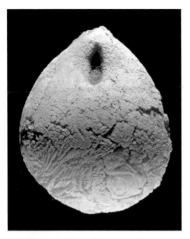

Fig. 493.   Fossil seed inner cast.

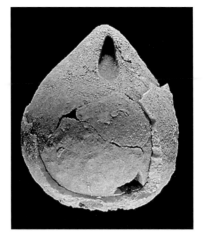

Fig. 494.   Fossil Seed with partial seed
coat.

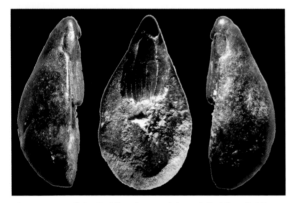

Fig. 495.   Fossil Seed with embryo. Left lateral, facial, and right
lateral views. Oldman Formation, Alberta.

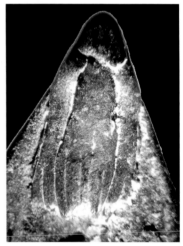

Fig. 496.   Close-up of embryo from Fig.
495.

Unfortunately, the embryo and lateral apexes are incomplete.

Could the developmental evolution of *Torreya* contain seeds with multiple developed embryos per ovule or is this just a case of an unusual preserved aberrant?

These seeds have not been studied.

Gnetaceaepollenites

The palynomorph form genera *Gnetaceaepollenites* from the formation (Fig. 498), although containing the Latin root name similar to the extant Gnetales, may not represent the order and is suspected by many to actually represent the Equisetaceae. As such, even if proven to be of the order, it is unknown what family would be represented. The Welwitschiaceae and Ephedraceae have fairly similar pollen when viewed under the light microscope while the Gnetaceae differs substantially.

These three families also live in very diverse habitats and show highly adaptive specializations. Welwitschiaceae are desert conifers restricted to the Namibian desert of Africa. Plants contain only two constantly growing leaves and large tap roots.

The Ephedraceae (Fig. 497) are shrub or small tree desert conifers of the northern hemisphere and South America. Their leaves are scale-like and whorled.

The Gnetaceae are evergreen shrubs or trees found in tropical forests of India. Until the palynomorphs can be more precisely identified, or plant remains are found, no attempt can be made to hazard a guess as to which family is represented, if any.

Fossil pollen:

Gnetaceaepollenites nobilis

Gnetaceaepollenites oreadis

Gnetaceaepollenites ventrosus

Fig. 497. *Ephedra nevadense*, mormon tea.

Fig. 498. *Gnetaceaepollenittes ventrosus* pollen.

Fig. 499. *Araucariacites australis* pollen.

Araucariacites australis

Although this pollen genus name suggests a connection with *Araucaria*, it is not typical *Araucaria*-like pollen (personal observation by D. Braman). The pollen grain is enigmatic (Fig. 499). Araucarian fossils have not been found in the formation.

Araucaria first occurs in the Jurassic (Miller 1988). In the Upper Cretaceous, *Araucaria* appear to have made it only as far north as Utah (Stockey 1994) in the central plains. Living *Araucaria* are now restricted to South America and Australasia.

Angiospermophyta

The Angiospermophyta are also called the flowering plants. This division is divided into three classes, the Magnoliopsida (Dicotyledonae), Eudicotyledonae, and Monocotyledonae. Both the Magnoliopsida and Eudicotyledonae are commonly referred to as dicotyledonous plants. *Dicotyledonous* refers to the seed containing two (di) seed leaves (cotyle). Dicots are also distinguished by normally having floral parts in groups of four or five (compare flowers; Figs. 500–507). Vascular bundles in their stems contain secondary growth restricted to the periphery.

The first remains of angiosperms are of suspected angiosperm-like pollen reported from the Triassic of Virginia (U.S.A.) and Jurassic of France (Zavada 2007). Angiosperm plant remains and palynomorphs show a major radiation during the mid-Cretaceous (131–91 mybp) (Crane et al. 2004).

Angiosperm dicotyledonous mega-fossil remains are rare when compared to gymnosperm remains in the formation. Petrified dicotyledonous wood is extremely rare. Presently only small, less than 1 cm in diameter, branch sections have been found in the Horseshoe Canyon Formation. This most likely reflects a preservational bias rather than a collection bias of these wood types based on random collections and sporadic active searches. The fluvial nature of deposition in the formation and the ability for modern dicotyledonous woods to undergo rapid decay pre- or post-burial may be a cause for absence of these fossil wood types as larger samples.

Angiosperm dicotyledonous wood samples with diameters of up to 3 cm have been recovered from the Campanian, Oldman Formation of Al-

Fig. 500. *Banksia menziesii*, firewood banksia.

Fig. 501. *Helianthus annuus*. wild sunflower.

Fig. 502. *Prunus cerasus*, sour cherry.

Fig. 503. *Asparagus densiflorus.*

Fig. 504. *Asarum (Hexastylis)* sp. wild asian ginger.

berta (Fig. 508) during searches for vertebrate remains. This may indicate a collection bias. Active searches for wood remains would most likely yield larger samples.

A variety of fossil angiosperm dicotyledonous woods of tree size, some with stem diameters of up to 1.3 metres, have been found in the Late Campanian to Early Maastrichtian of Texas (Lehman and Wheeler 2001; Lehman et al. 1994) during active searches.

Angiosperm wood specimens from the Upper Cretaceous (Cenomanian to early Campanian) of Antarctica also have large stem diameters up to 44 cm (Poole and Cantrill 2001).

Although a few dicotyledonous leaf beds are known in the Horseshoe Canyon Formation, most contain a reduced assemblage representing deciduous trees from the immediately surrounding area or location (Fig. 509). The remains of evergreen angiosperms at present have not been identified in the flora, although they most likely occurred. It is also very unfortunate that, although present, none of the leaf remains have been formally studied or described since W. A. Bell's 1949 studies. This guide presently relies mainly on fossil fruits, seeds, and palynomorphs that highlights the obvious gap in the knowledge of leaf identification from the formation. The area of leaf identification is in much need of study.

Fig. 505. *Castilleja* sp. indian paint brush.

Fig. 506. *Epiphyllum* sp. hybrid. epiphytic cactus.

Fig. 507. *Hydrangea paniculata* 'grandiflora.' snowball hydrangea.

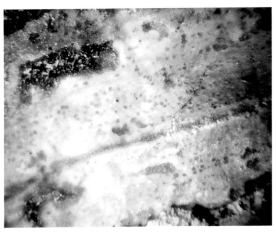

Fig. 508. Close-up of unprepared fossil angiosperm wood, Oldman Formation.

Fig. 509. Fossil leaf.

Fig. 510. *Sassafras albidum* trilobed leaf.

Fig. 511. *Sassafras albidum* single entire leaf. From same branch as Fig. 510.

MAGNOLIOPSIDA (DICOTYLEDONAE)

Laurales

LAURACEAE (LAUREL FAMILY)

Sassafras-like

The earliest occurrence in the fossil record of *Sassafras* is in the Santonian/Campanian and Maastrichtian of Antarctica in the form of wood (Poole et al. 2000). The next occurrence, in the form of leaves, is in the Paleocene, Fort Union Formation of northwestern United States (Tidwell 1998). Although Lauraceae pollen has not been identified in the Horseshoe Canyon Formation, seeds tentatively identified as *Sassafras*-like have been found.

Fossil seeds (Figs. 513–515) from the Horseshoe Canyon Formation are up to 1 cm in diameter, round to oblong with a raised lateral keel. The apex, in well-preserved specimens, contains two lateral vascular cylinders (Fig. 515) with a depressed seed-attachment scar. Basally there is a small laterally compressed point.

These seeds, which are common, are sometimes referred to as fossil cherry pits by collectors, but do not appear related to *Prunus*.

Living *Sassafras* contains three species: *S. tzumu* in central China, *S. randaiense* (sometimes classified as *Yushunia randaiensis*) in Taiwan, and *S. albidum* (Figs. 510–512) from the eastern United States. Living *Sassafras* leaves are polymorphic and range from oblong elliptical to apically three-lobed. *Sassafras* is a deciduous medium-sized tree or shrub that prefers rich acidic soil in shaded areas.

Fig. 512. *Sassafras albidum* Flowers.

Fig. 513. Fossil seed.

Fig. 514. Fossil seed.

Fig. 515. Fossil fractured seed showing apex and two vascular strands.

Fig. 516. *Nymphaea* sp.

Fig. 517. *Nymphaea* sp. ventral view of leaf showing venation.

Nymphaeales

NYMPHAEACEAE (LOTUS FAMILY)

Presently the earliest confirmed fossil evidence for Nymphaeaceae are flowers from the Turonian (90 mybp) of North America (Crepet et al. 2004).

Although leaves or other plant parts have not yet been identified in the Horseshoe Canyon Formation (Figs. 516, 517), pollen representing the Nymphaeaceae is preserved (Fig. 518). With this information it can be stated that water plants referable to the Nymphaeaceae may be present in the flora.

In the extant family are six genera with up to fifty-eight species.

Fossil pollen:
Zonosulcites scollardensis

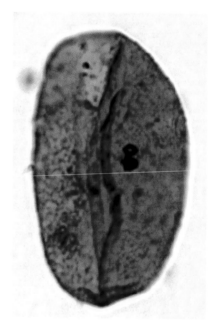

Fig. 518. *Zonosulcites scollardensis.*

Ceratophyllales

CERATOPHYLLACEAE (HORNWORTS)

Ceratophyllum

The oldest confirmed record of the Ceratophyllaceae is based on achenes from the Paleocene of North America (Herendeen et al. 1990).

The highly diagnostic achenes of *Ceratophyllum* have recently been found in the Horseshoe Canyon Formation. Achenes are ellipsoidal and up to 5 mm in length (Figs. 519, 520). In transverse section the achene body is lenticular in view. The achene surface is covered with the remnants of surface spines. Based on the basal spine remnants, the apical spine appears to have been more pronounced. These achenes have not been formally described.

Six species of *Ceratophyllum* occur worldwide (Les 1989) and range from temperate to tropical. All are submerged or floating, deciduous, perennial water plants which spread rapidly and prefer cool water habitats. Plants contain an axis with whorled finely divided leaves. Pollination occurs underwater.

Ceratophyllum submersum bears warty but spineless achenes and occurs from Europe to Africa and Asia. *Ceratophyllum demersum* from North America bears achenes that are ellipsoidal, up to 5 mm long and contain a spine-like beak and two spines near the base. Achenes of *Ceratophyllum echinatum* from the southeastern United States and northern Mexico are also up to 5 mm long and have a roughened surface as well as basal and lateral spines.

The pollen of living *Ceratophyllum* is small, thin-walled and non-porate. It has not been described from the Cretaceous fossil record to date. This may be due to its very non-descript nature or poor potential to fossilize due to its thin wall.

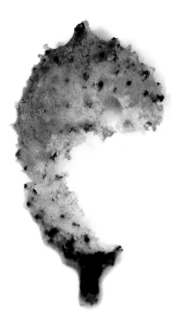

Fig. 519.   Silicified hollow cast of achene.

Fig. 520.   Silicified achene.

Fig. 521. *Cercidiphyllum japonicum* foliage.

Fig. 522. *Cercidiphyllum japonicum* flowers.

Fig. 523. *Cercidiphyllum japonicum* leaf.

EUDICOTYLEDONAE

The Eudicotyledons represent a major group of non-magnolid dicots. A criterion for being placed in the Eudicots is the presence of triaperaturate or triaperaturate-derived pollen. Eudicot pollen first occurs in the Barremian, mid-Cretaceous (126 mybp).

Saxifragales

CERCIDIPHYLLACEAE (KATSURA)

The Cercidiphyllaceae is represented in the formation by pollen (Fig. 527), leaf imprints (Fig. 524) as well as calcified, silicified, and carbonized fruits and stems. Only the pollen has been formally described.

In the past, finds of this kind in the formation were placed in the form genus *Trochodendroides arctica* for the leaves and *Jenkinsella arctica* for the seed pods (Bell 1949).

Seed pods up to 1.6 cm in length can be found as silica replaced roll outs from ironstones (Fig. 526). When etched from the rocks, seed pods are grouped on branches and have very characteristic morphology (Fig. 525), which can also be seen in carbon-trace specimens.

Many Upper Cretaceous and Tertiary genera and species of Cercidiphyllaceae have been reported from a variety of formations around the world (Krassilov 1976; Crane 1984; Crane and Stockey 1985). Some sites have produced more than one species, but these are felt to most likely represent variations of organs in a single species (Crane and Stockey 1985).

Do the many Upper Cretaceous and Tertiary genera and species actually reflect differing forms or do they merely reflect a geological distribution of a very much species-reduced plant evolving through time similar to *Parataxodium* and *Metasequoia*?

Genera such as *Joffrea speirsii* (Crane and Stockey 1985), although thought not to be a plausible direct ancestor to living *Cercidiphyllum* (Crane and Stockey 1985), may in fact be a direct link. The changes in phyllotaxy between these two genera are no greater than that of *Parataxodium* and *Metasequoia*. The changes in the number and orientation of the follicles appear minor when one considers 60+ million years or evolution to the extant form.

Living *Cercidiphyllum* is monotypic with *Cercidiphyllum japonicum* (Figs. 521–523), the Katsura, from Japan and *C. japonicum* var. *sinensis* from central and western China. The plant is deciduous and sometimes up to 30 m tall, but it is most commonly shrub-like up to 5 m tall and 3 m wide. It prefers rich moist soils in full sun or partial shade. Katsura foliage emits a pungent Fall aroma reminiscent of burnt toffee.

Cercidiphyllum japonicum has been grown as a shrub (3 m tall) in the town of Drumheller for ten years by the author.

Fossil pollen:
Circumflexipollis tiliodes

Fig. 524.   Fossil leaf.

Fig. 525.   Silicified seed pods.

Fig. 526.   Silicified seed pod.

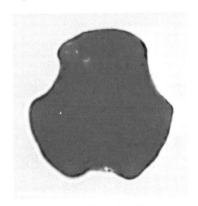

Fig. 527.   *Circumflexipollis teiliodes* pollen.

HALORAGACEAE

Myriophyllum-like

The family Haloragaceae contains genera such as *Proserpinaca, Vinkia, Glischrocaryon, Meziella, Haloragodendron, Haloragis,* and the well-known *Myriophyllum* that contains up to fifty species, ranging from warm temperate to tropical.

An aquatic *Myriophyllum*-like plant has been found in the formation. The fossil consists of diagnostic stem sections up to 3.1 mm in diameter. Sections contain differing states of preservation (Figs. 528–534) and are comparable to living *Myriophyllum* (Fig. 535).

The Cretaceous (70 mybp) fossil *Obispocaulis myriophylloides* from Mexico (Hernández-Castillo and Cevallos-Ferriz 1999) has morphology similar to the Drumheller fossil.

A palynomorph for this megaplant is unknown in the formation. The fossil remains undescribed.

Fig. 528. Fossil Myriophyllum sp. cross-section.

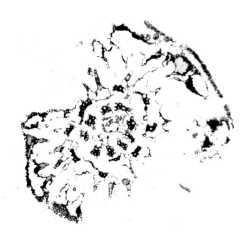

Fig. 529. Fossil Myriophyllum sp. cross-section showing differential preservation.

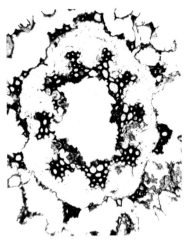

Fig. 530.   Close-up of stele in Fig. 528.

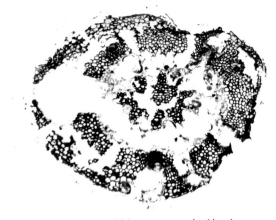

Fig. 531.   Cross-section with better preserved epidermis.

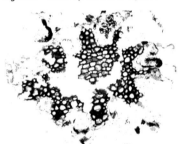

Fig. 532.   Close-up of stele in Fig. 531.

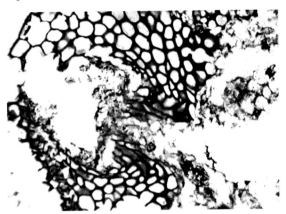

Fig. 533.   Close-up of vascular departure in Fig. 531.

Fig. 534.   Close-up of air lacunae in Fig. 528.

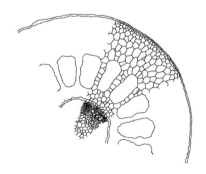

Fig. 535.   Drawing of cross-section of Myriophyllum brasiliense. (redrawn and adapted from Ogden 1974).

Fig. 536. *Hamamelis virginiana* leaf.

HAMAMELIDACEAE (WITCH HAZELS)

The pollen of the Hamamelidaceae from the formation are grouped with the Platanaceae due to their close morphological similarities, which makes them difficult to separate.

Presently the fossil record of the Hamamelidaceae in the Horseshoe Canyon consists only of pollen. Fossilized pistillate inflorescences, fruits, and pollen grains are known from the Cretaceous, Turonian (90 million years) of North America (Zhou et al. 2001).

Approximately twenty-seven genera and up to eighty-two extant species are known with the majority of plants represented by deciduous small trees or shrubs. The genus *Hamamelis* contains *Hamamelis virginiana* (Fig. 536) from the eastern United States, *H. vernalis* from Missouri and Louisiana, *H. macrophylla* from Georgia and Texas, *H. mollis* from Hubei and Kiangsi provinces in China, and *H. japonica* from the mountains of Japan.

Proteales

PLATANACEAE (PLANE TREE)

The Platanaceae earliest known occurrence is in the Cenomanian of Bohemia and Kansas (Manchester 1999). The Platanaceae are represented in the Horseshoe Canyon Formation by calcified and partially silicified seed heads and carbon-trace leaf fossils. Fossil leaves of the Platanaceae are comparatively large (Fig. 538) and range in morphologies from simple to palmate to pinnate. Some leaves belonging to the Platanaceae in the Cretaceous have been erroneously identified as *Vitis stantoni*.

Although several extinct genera exist in the fossil record, the Platanaceae today consists of the single genus *Platanus*. The genus is divided into two subgenera: *Platanus* (Fig. 537) with seven species and *Castaneophyllum* with one species. Species of *Platanus* occur in North America, Europe, Crete, Asia, and the Himalayas.

All represent deciduous trees that require full sun in deep well-drained soils.

Fossil Platanaceae/Hamamelidaceae pollen:
Tricolpites micromunus
Tricolpites mtchedlishviliae
Tricolpites occidentalis
Tricolpites parvus
Tricolpites vulgaris
Tricolpopollenites prolatus

Fig. 537. *Platanus occidentalis* leaf.

Fig. 538. Large fossil leaf attributed to *Platanus*.

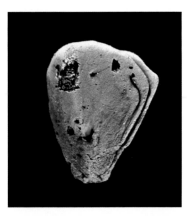

Fig. 539.   Fossil fruitlet showing folded coat.

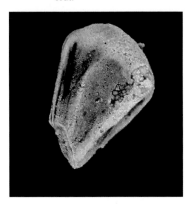

Fig. 540.   Fossil fruitlet showing small apical spine.

Trochodendrales

TROCHOCENDRACEAE (WHEEL TREE)

Nordenskioldia-like

The Trochodendraceae first appears in the middle Cretaceous of Asia and North America (Crane 1989; Manchester 1999). Many Cretaceous leaf fossils assigned to the Cercidiphyllaceae may in fact belong to this family. The family contains three genera: *Trochodendron* with one extant species; *Tetracentron* with one extant species; and *Nordenskioldia*, which represents an extinct Tertiary genus.

Silicified and calcite-replaced fruits reminiscent of *Nordenskioldia* are found in the formation. Although reminiscent of *Nordenskioldia* in containing sessile fruits composed of many single-seeded wedge-shaped fruitlets arranged in a whorl, there are a few major differences.

The Cretaceous fossil consists of compacted fruits on a highly reduced infructescence. This differs from the tertiary forms with their elongate infructescences (Fig. 546) and the persistent umbrella-like structure after seed dispersal of Tertiary forms (Manchester et al. 1991). Each Cretaceous fruit consists of a well-packed whorl of up to nineteen fruitlets attached to a reduced receptacle or cup-like base (Figs. 543–545). Individual fruitlets are wedge-shaped, up to 6 mm long, and contain a re-curved persistent style, which varies in length depending on the fruitlet position and/or maturation (Figs. 539–543). Fruitlets are interpreted as being shed individually.

This fossil has not been named but is described in Serbet (1997) as "Trochodendraceous infructescence axes, seed morphology and anatomy" as well as seed type G.

This Cretaceous fossil is in the Trochodendraceae and represents the early plant evolution to the genus *Nordenskioldia*. *Nordenskioldia* became extinct in the Miocene.

The nearest living representative, *Trochodendron aralioides*, is an evergreen tree from the mountains of Japan, southern South Korea, and Taiwan.

Fossil pollen of *Nordenskioldia* has not been identified in the formation.

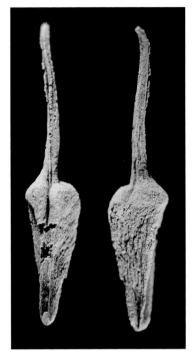

Fig. 541.   Two views of a fossil fruitlet with pronounced spine.

Fig. 542.   Three fossil fruitlets together.

Fig. 543.   Two views of same fossil infructescence with retained partial fruits and fruitlets.

Fig. 544. Large fossil infructescence with shed fruitlet scars.

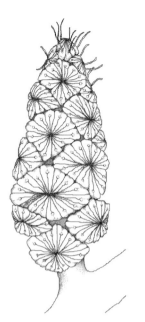

Fig. 545. Drawing of Cretaceous infructescence attached to branch.

Fig. 546. Line drawing of Miocene infructescence. *Nordenskioldia interglacialis* based on Miocene specimen UWBM 57254. Drawn from Manchester et al. (1991).

Rosales

ULMACEAE (ELM FAMILY)

The Elm family contains six genera and up to thirty-five species. These are commonly deciduous trees or shrubs found throughout the northern hemisphere (Figs. 547, 548). The earliest occurrence of the Ulmaceae in North America is from the Paleocene (Manchester 1999). There is only one species of palynomorph in the Horseshoe Canyon Formation representing the family (Fig. 549).

Fossil pollen:
Ulmipollenites planeraeformis

Fig. 547. *Ulmus pumila*, siberian elm, seeds.

Fig. 548. *Ulmus pumila*, siberian elm, leaf.

Fig. 549. *Ulmipollenites planeraeformis* pollen.

Rhamnaceae (Buckthorn)

Fig. 550. *Rhamnacidites minutapollenites* pollen.

The Rhamnaceae consist of fifty-five genera and up to nine hundred species. The family consists chiefly of tendril-bearing vines that are cosmopolitan in distribution.

The fossil record in the Horseshoe Canyon Formation consists of pollen and possible leaves. Only the pollen has been formally described to date (Fig. 550).

Fossil pollen:

Rhamnacidites minutapollenites

Fagales

MYRICACEAE (MYRTLES, BAYBERRY)

This family consists of three genera and up to fifty-seven species of deciduous to evergreen shrubs or small trees that range from temperate to subtropical climates. Plants occur in North America, southeast Asia, and Japan. The majority of myrtles prefer moist to saturated soils with some species found in bogs or swamps.

Plant remains have not been found and as such the fossil record in the formation consists only of pollen to date. The Myricaceae also occur in the Eocene of North America (Manchester 1999).

Fossil pollen:
Myricipites dubius

FAGACEAE (BEECHES)

The beeches contain seven genera with up to 670 species. Palynomorphs in the formation are suggestive of the genera *Fagus* and *Quercus* (Fig. 551). *Fagus* are tall, temperate, deciduous trees represented by ten living species.

Quercus or Oaks have up to five hundred recent species and are found in North America, northwestern South America, temperate and subtropical Eurasia, and northern Africa. *Quercus* and *Fagus* contain both evergreen and deciduous forms.

Although pollen appears in the formation, plant remains do not.

Fossil pollen:
Faguspollenites granulatus
Quercoidites sternbergii

Fig. 551. *Quercus macrocarpa* leaf.

Fig. 552. *Alnus glutinosa* european alder leaf.

Fig. 553. *Alnus glutinosa* seed cones. Left with seed; right after seed shedding.

Fig. 554. *Alnipollenites finitimus* pollen.

BETULACEAE (BIRCHES)

Presently two genera of palynomorphs are recognized from the Horseshoe Canyon Formation with the first being *Alnus*-like (Fig. 554) and the second *Carpinus*-like.

The Betulaceae consists of six genera: *Betula, Alnus, Corylus, Carpinus, Ostryopsis,* and *Ostrya. Betula* represents up to sixty species of deciduous trees and shrubs (Figs. 555, 556) that range from the northern temperate to arctic regions. *Corylus* consists of fifteen species of large deciduous shrubs (Fig. 557) found in temperate North America, Europe, and Asia. *Alnus* has up to thirty species of deciduous temperate trees (Figs. 552, 553) in the northern hemisphere and extends into the Andes of South America. *Ostrya* is deciduous and contains about ten species in the northern hemisphere.

Carpinus-like

The *Carpinus*-like genus is represented by silicified nutlets and pollen in the formation. Nutlets are small, up to 3 mm in length, and teardrop shaped (Figs. 559, 560). The nutlet contains a convex outer face, which is characteristically ribbed while the inner aspect is somewhat flat and smooth without ribbing.

Carpinus-like nutlets and leaves assigned to *Palaeocarpinus* (Crane 1981) are widespread in Paleocene strata (Manchester 1999). It is unknown if the Cretaceous nutlet contained the spiny bracts associated with Paleocene forms (Crane 1981, Manchester and Guo 1996, Manchester and Chen 1996); therefore, it cannot be placed with certainty in *Palaeocarpinus* at this time. Nutlets identifiable as belonging to *Carpinus* occur in the late Eocene of China and middle Eocene of Europe and North America (Manchester 1999).

In the genus *Carpinus*, there are two subgenera recognized *Carpinus* and *Distegocarpus*, with up to sixty species distributed in three major areas: Europe, Asia, and North America (Fig. 558). The fossil nutlets, based on a limited search, are most similar to *C. caroliniana*, the American Hornbeam. *C. caroliniana* from eastern North America is up to 10 m tall, monoecious, and deciduous.

Fossil pollen:
Alnipollenites finitimus
Alnipollenites speciipites
Carpinites ancipites

Fig. 555. *Betula papyrifera*, paper birch, leaf.

Fig. 556. *Betula papyrifera* seed cones. Left with seed; right after seeds and bracts (below) shed.

Fig. 557. *Corylus Americana*, hazelnut leaf.

Fig. 558.   *Carpinus turczaninovii*, leaf.

Fig. 559.   A fossil *Carpinus* sp. nutlet showing the outer convex face and ribs.

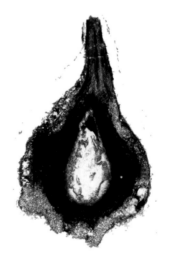

Fig. 560.   Longitudinal section of a fossil nutlet.

Malpighiales

The Malpighiales earliest known occurrence is in the Turonian of New Jersey (Crepet and Nixon 1998).

SALICACEAE (WILLOW FAMILY)

The Salicaceae contains fifty-five genera with approximately 1,010 species. They are dioecious, deciduous trees and shrubs with an almost world-wide distribution (Fig. 561).

Only fossil pollen presently represents this family in the formation (Fig. 562).

Fossil pollen:
Rousea subtilis

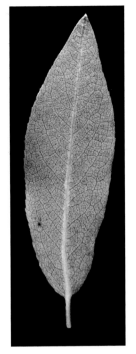

Fig. 561.   *Salix amygdaloides* leaf.

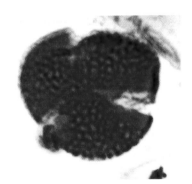

Fig. 562.   *Rousea subtilis* pollen.

Ericales

SYMPLOCACEAE (SWEETLEAF FAMILY)

Fig. 563. *Symplocoipollenites morrinensis* pollen.

The Symplocaceae consists of one genus with about 320 species of trees and shrubs. They are native to warm temperate and tropical areas of Eurasia, Australia, and North America. *Symplocos* occur in seasonally wet woodlands in acidic soils. The fossil record in the formation consists only of one form of pollen (Fig. 563).

Fossil pollen:
Symplocoipollenites morrinensis

Caryophyllales

AMARANTHACEAE

The Amaranthaceae is a large family with approximately 174 genera and 2,050 to 2,500 species of herbs and shrubs. The family is only represented by one fossil pollen type (Fig. 564) in the formation, so it is presently impossible to guess its generic plant affinity.

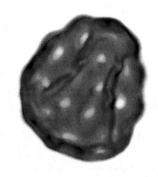

Fossil pollen:
Polyporina cribraria

Fig. 564. *Polyporina cribraria* pollen.

Fig. 565.   *Gunnera manicata* flower.

Gunnerales

GUNNERACEAE

The pollen form *Tricolpites reticulates* (Fig. 566) has been identified as belonging to the genus *Gunnera* in various formations. Presently *Gunnera* (Fig. 565), which contains 40–50 species, ranges in size from 40 cm to 2.5 m high. It would be of interest to see which plant form the fossil pollen possibly represents.

Fossil pollen:
Tricolpites reticulatus
Tricolpites microreticulatus

Fig. 566.   *Tricolpites reticulatus* pollen.

Myrtales

LYTHRACEAE

Nymphaeites angulatus

In the Horseshoe Canyon Formation, dispersed leaves assignable to the Trapaceae are known. These leaves were identified as *Nymphaeites angulatus* from the formation (Bell 1949). This plant has undergone many name changes over time in other formations such as *Neuropteris angulata*, *Trapa microphylla*, *Trapa cuneata*, *Nymphaeites angulatus*, *Trapa angulatus*, and *Quereuxia angulata*.

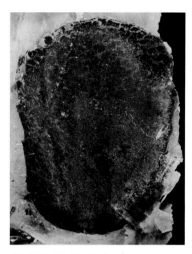

Fig. 567.   Silicified leaf.

Fossil leaves are obovate and contain serrations with double mucronate tips, which are orientated front to back with a median sinus or gland between them. They have been found preserved as both carbon trace and silicified (Figs. 567, 569).

The fossil leaves are identical to those found in the Paleocene of Saskatchewan described as *Trapago angulata* (McIver and Basinger 1993).

In the Alberta Cretaceous, a correlative formation to the Horseshoe Canyon Formation is the St. Mary River Formation. In the St. Mary River Formation complete vegetative plants named *Trapago angulata*, previously figured by W. A. Bell (1949) as *Nymphaeites angulatus*, are known (Stockey and Rothwell 1997). Leaves of this plant are identical to the silicified fossils from Drumheller. Fossil rhizome sections also appear identical in form (Fig. 568).

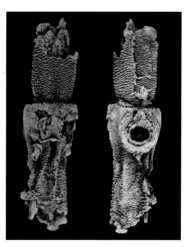

Fig. 568.   Two views of the same fossil silicified stem section.

Irrespective of which genus name is considered valid at the present time, debate still lingers on the familial placement of this fossil. New information supplied by silicified leaves from this formation gives greater credence to its placement in the Trapaceae. The question as to the teeth on the

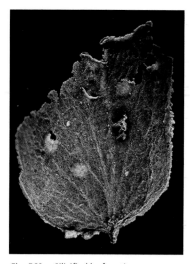

Fig. 569.   Silicified leaf section.

leaves having a double mucronate tip is confirmed with silicified material from the formation.

These *Trapa*-like plants, which are found in the Upper Cretaceous and Paleocene of North America and eastern Russia, represent an extinct genus in the Trapaceae.

Trapa is represented by up to five living species, all occurring in China, Taiwan, and Japan, with some species introduced into the southeastern United States. *Trapa* are perennial or annual, deciduous floating water plants that prefer shallow water or muddy soils. *Trapa* is sometimes placed in the Onagraceae.

Fossil pollen is unknown.

NYSSACEAE (NYSSA)

The Nyssaceae consists of up to twenty-two species from five genera: *Diplopanax, Mastixia, Davidia, Camptotheca,* and *Nyssa.*

Nyssa-like

Fossil fruit tentatively identified as *Nyssa*-like have been found at various stages of development. These fruit are linear, up to 1.2 cm long, and triangular in cross-section (Figs. 572–574). Three ridges on the widest face commonly extend the length of the fruit body in non-abraded specimens. Internally they contain one to three locules (Fig. 574) without a central axial bundle. Whole fruits are found as well as fruits missing 1/3 of the upper section of a single lateral wall (Fig. 573). Fruits degrade further into cross-section pieces.

Fossil fruits are un-named seed types D, E, I, J, L, and O of Serbet (1997).

The oldest confirmed occurrence of *Nyssa* in the fossil record is from the Eocene of North America (Manchester 1999) based on diagnostic stones with dorsal, apical germination valves. The fossil from Drumheller also appears similar to the Paleocene fossil *Amersinia* (Manchester et al. 1999). It is unknown if the Cretaceous fossil contains the cycle of prominent bract scars seen in *Amersinia*. Both *Nyssa* and *Amersinia* appear closely related.

Nyssa, called "Tupelo" or "Sour Gum" (Figs. 570, 571), consists of four species in North America and two in China. All *Nyssa* are deciduous and prefer river basins and standing water about their roots for part of the year.

Fig. 570. *Nyssa sinensis* leaf.

Fig. 571. *Nyssa sinensis* flower.

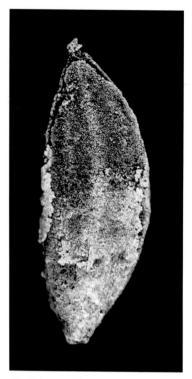

Fig. 572.   Fossil fruit.

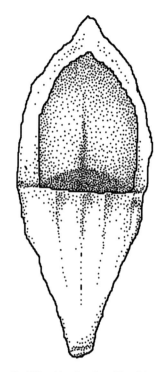

Fig. 573.   Line drawing of fossil fruit with germinal valve missing. Drawn from Serbet (1997).

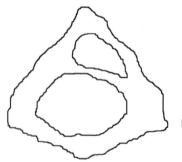

Fig. 574.   Line drawings of fossil fruit cross-sections containing 2 or 3 ovule cavities. Drawn from Serbet (1997).

Davidia is a monotypic deciduous tree up to 20 m tall found in China (Fig. 575). *Davidia involucrata*, the Dove Tree, bears characteristic long-stalked drupes, which are very diagnostic. Fossil *Davidia* is known from drupes up to 18 mm long (Figs. 577–579) with up to seven ovule cavities (Fig. 580) and possible leaves (Fig. 576) from the formation. Drupes were designated as seed types C, M, and N (Serbet 1997). These fossils represent the oldest occurrence of the genus.

Fossil *Davidia* occurs in the Paleocene of eastern Russia and North America (Manchester 2002).

Four pollen forms representing the Nyssaceae are found in the formation (Fig. 581).

Fossil pollen:
Nyssapollenites bindae
Nyssapollenites parvus
Nyssoidites edmontonensis
Nyssoidites ingentipollinus

Fig. 575.   *Davidia involucrata* leaf.

Fig. 576.   Possible fossil *Davidia*-like leaf.

Fig. 577. Fossil drupe with remnants of fleshy outer coat preserved between ribs.

Fig. 578. Fossil drupe showing woody ribs.

Fig. 579. Fossil drupe with woody ribs eroded at ends.

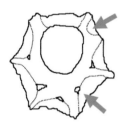 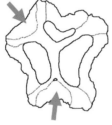  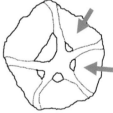

Fig. 580. Line drawings of fossil drupe cross-sections showing ovule cavities and valves (stippling). Note the many abortive ovule cavities (some with arrows). Drawn from Serbet (1997).

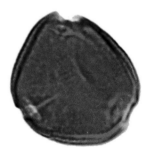

Fig. 581. *Nyssapollenites* sp. pollen.

Santalales

LORANTHACEAE/SANTALACEAE

The Loranthaceae and the also sometimes included Santalaceae consist of up to 109 genera and 2,940 species of warm temperate to tropical semi-parasitic to parasitic plants. Many are genera specific to a specific host such as Pine Mistletoe (Fig. 582), which prefers to parasitize immature Lodgepole Pine. Some Australian mistletoe species prefer to parasitize deciduous and non-deciduous trees in partial shade but avoid oaks and conifers. All species are considered pests, although one or two per tree causes little damage.

The common mistletoe, *Viscum album*, from Europe and Asia is evergreen with a spread of up to three metres and is considered warm temperate.

Presently the fossil record from the formation consists only of pollen (Fig. 583).

Fossil pollen:
Cranwellia bacata
Cranwellia edmontonensis
Cranwellia asperata
Cranwellia rumseyensis
Cranwellia striata
Loranthacites pilatus
Loranthacites catterallii

Fig. 582.   *Arceuthobium americanum*, pine mistletoe.

Fig. 583.   *Cranwellia rumseyensis* pollen.

Fig. 584. *Ilex deciduosa* leaf.

Aquifoliales

AQUIFOLIACEAE

Ilex-like

The Aquifoliaceae consists of one genus with up to 405 species of both evergreen (Figs. 585, 586) and deciduous (Fig. 584) trees and shrubs found in Europe, North Africa, and Asia.

Seeds have been extracted from ironstones in the Horseshoe Canyon Formation which appear *Ilex*-like. The seeds are up to 16 mm long and wedge-shaped with ribbing on the surfaces. A drawing (Fig. 587) is given here based on a seed type B (Serbet 1997). These appear visually similar to those produced by *Ilex deciduosa*. Pollen from the family is found in the formation.

Fossil pollen:
Ilexpollenites obscuricostatus

Fig. 585. *Ilex vomitoria* leaf.

Fig. 586. *Ilex vomitoria* flowers.

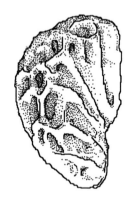

Fig. 587. Fossil seed. Drawn from Serbet (1997).

Buxales

BUXACEAE (BOX)

The Buxaceae consists of four genera and approximately seventy species of warm temperate to tropical dioecious and monoecious herbs, shrubs, and trees with almost worldwide distribution (Fig. 588). The fossil record from the formation to date consists of pollen only (Fig. 589).

Fossil pollen:
Erdtmanipollis albertensis
Erdtmanipollis procumbentiformis
Erdtmanipollis cretacea

Fig. 588.   *Sarcococca humilis* flowers and fruit.

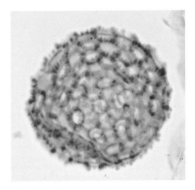

Fig. 589.   *Erdtmanipollis procumbentiformis* pollen.

Fig. 590. *Rhus aromatica* leaf.

Fig. 591. *Ailanthipites cupidineus* pollen.

Sapindales

ANACARDIACEAE (CASHEW FAMILY)

Anacardiaceae pollen found in the formation was used in the past as an indicator of tropical conditions (Srivastava 1970) as it contains the well-known cashew nuts. The family contains seventy genera and 985 species. Some well-known warm temperate genera such as *Rhus* (Sumacs) (Fig. 590) occur in both the China and North America. Both plant groups would represent different climatic regions. Unfortunately neither plant form has been found as a macrofossil in the formation.

Fossil pollen:
Rhoipites crassus

SIMAROUBACEAE (QUASSIA)

The family contains nineteen genera with ninety-five species of mostly subtropical to tropical trees and shrubs. The palynomorph name (Fig. 591) hints at the genus *Ailanthus*, which is represented by the living monotypic species *Ailanthus altissima* or Tree of Heaven, although none of the characteristic samaras or leaves have been found or identified in the formation.

The fossil record of the Simaroubaceae extends only as far as the Eocene of North America, Asia, and Europe (Manchester 1999).

Ailanthus is a medium-sized tree from Central China, which can survive poor soils and air pollutants.

Fossil pollen:
Ailanthipites cupidineus
Ailanthipites potoniei

Apiales (Umbellales)

TORICELLIACEAE

Toricellia-like

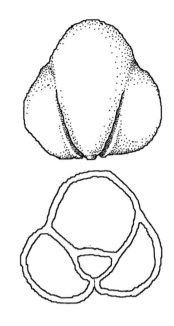

Only one seed form in the formation has been found that appears to belong to the family.

Seeds of the Toricelliaceae, which show diagnostic characteristics of *Toricellia*, are extremely rare and are found only at one site. Seeds are up to 2 mm long with a central body and two laterally inflated wings (Fig. 592). It is not known if one of the unknown pollens listed also represents this genus. The seeds were identified as unknown seed type F (Serbet 1997).

The earliest record of the genus *Toricellia* based on seeds are from the Eocene of North America and the Miocene of Austria (Manchester 1999).

Fig. 592. Drawing of seed figured by Serbet (1997). External drawing (top), cross-section line drawing (bottom).

Fig. 593. *Fraxinus.* sp. leaf.

Lamiales

OLEACEAE (OLIVE)

The Oleaceae contains about twenty-four genera and 615 species of vines, trees, or shrubs from temperate to tropical regions of both hemispheres. The palynomorph from the formation seems to have morphology similar to the genus *Fraxinus* (Fig. 593). Unfortunately only a single palynomorph is known (Fig. 594) without any mega-fossil remains.

Fossil pollen:
Fraxinoipollenites constrictus

Fig. 594. *Fraxinoipollenites constrictes* pollen.

Incertae Sedis

Palynomorphs

Many palynomorphs are named from the formation but their affinities are unknown. These are broken down into three headings: A Unique Palynomorph; Extinct Triprojectates; Unknown Angiosperms.

A Unique Palynomorph

Fig. 595. *Kurtzipites andersonii* pollen.

Although found only as a palynomorph in Alberta Cretaceous sediments, *Kurtzipites* pollen (Fig. 595) in the form of *K. trispissatus* has been found attached to wholly staminate flowers in the Paleocene of Saskatchewan. The flowers are described as diclinous, anemophilous, and deciduous (McIver and Basinger 1993). Unfortunately the family from which these flowers would have come from is not identified at this time.

Fossil pollen:
Kurtzipites andersonii
Kurtzipites trispissatus

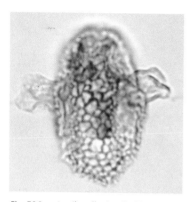

Fig. 596. *Aquilapollenites funkhouseri* pollen.

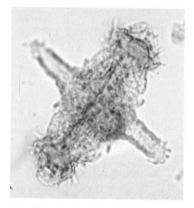

Fig. 597. *Aquilapollenites sentus* pollen.

Fig. 598. *Aquilapollenites senonicus* pollen.

Extinct Triprojectates

Many Aquila palynomorph forms are known from the formation. These pollen forms are all thought to perhaps belong to the Apiaceae, although this is far from certain.

Following is a listing of extinct Triprojectate pollen. A few palynomorphs are figured to show variation in shape and form (Figs. 596–600).

Aquilapollenites accipiteris
Aquilapollenites alveolatus
Aquilapollenites amicus
Aquilapollenites amplus
Aquilapollenites amygdaloides
Aquilapollenites antigonei
Aquilapollenites aptus
Aquilapollenites argutus
Aquilapollenites ascriptivus
Aquilapollenites asper
Aquilapollenites attenuatus
Aquilapollenites aucellatus
Aquilapollenites augustus
Aquilapollenites catenireticulatus
Aquilapollenites ceriocorpus
Aquilapollenites clarireticulatus
Aquilapollenites comosus
Aquilapollenites debilis
Aquilapollenites decorus
Aquilapollenites dispositus
Aquilapollenites dolium
Aquilapollenites drumhellerensis
Aquilapollenites firmus
Aquilapollenites funkhouseri
Aquilapollenites fusiformis
Aquilapollenites granulatus
Aquilapollenites hirsutus
Aquilapollenites hispidus

Aquilapollenites insignis
Aquilapollenites leucocephalus
Aquilapollenites macgregorii
Aquilapollenites medeis
Aquilapollenites minutes
Aquilapollenites mirabilis
Aquilapollenites oblatus
Aquilapollenites paplionis
Aquilapollenites petasus
Aquilapollenites polaris
Aquilapollenites cf. *A. procerus*
Aquilapollenites psilatus
Aquilapollenites pudicus
Aquilapollenites pumilis
Aquilapollenites quadrilobus
Aquilapollenites reductus
Aquilapollenites regalis
Aquilapollenites scabridus
Aquilapollenites senonicus
Aquilapollenites sentus
Aquilapollenites spinulosus
Aquilapollenites stelckii
Aquilapollenites stellatus
Aquilapollenites trialatus
Aquilapollenites validus
Aquilapollenites venustus
Aquilapollenites vinosus
Aquilapollenites sp.
Fibulapollis scabratus
Mancicorpus albertensis
Mancicorpus anchoriforme
Mancicorpus borealis
Mancicorpus calvus
Mancicorpus canadiana
Mancicorpus gibbus
Mancicorpus rostratus
Mancicorpus senonicus
Mancicorpus solidus

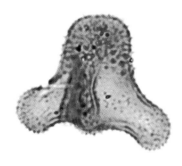

Fig. 599. *Aquilapollenites quadrilobus* pollen.

Fig. 600. *Orbiculapollis globosus* pollen.

Fig. 601. *Gunnaripollis superbus* pollen.

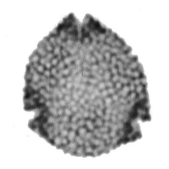

Fig. 602. *Proteacidites* sp. pollen.

Fig. 603. *Rosannia manika* pollen.

Mancicorpus tripodiformis
Mancicorpus vancampoi
Orbiculapollis globosus
Scollardia nortonii
Scollardia steevesii
Scollardia trapaformis
Translucentipollis plicatilis

Unknown Dicotyledonous Angiosperms

Many dicotyledonous angiosperm palynomorphs have not as yet been identified to family and, as such, are listed here as unknowns. Although some genera may have names that allude to a family or genus, the palynomorph does not appear to represent the plant upon closer examination. A few of the various forms are shown in figures 601–603.

Unknown Palynomorphs:
Accuratipollis evanidus
Albertispollenites rosalindii
Alnipollenites trina
Alnipollenites cf. *A. verus*
Beaupreaidites angulatus
Beaupreaidites libitus
Beaupreaidites occulatus
Beaupreaidites elegansiformis
Beaupreaidites sp.
Callistopollenites comis
Callistopollenites radiatostriatus
Callistopollenites tumidoporus
Disulcites magnus
Grewipollenites canadensis
Gunnaripollis superbus
Inaperturopollenites dubius
Liquidambarpollenites sp.
Loranthacites catterallii
Marcellopites basilicus

Marcellopites tolmanensis
Margocolporites taylorii
Monosulcites drumhellerensis
Monosulcites echinatus
Monosulcites gemmatus
Monosulcites sp.
Nothopollenites primus
Polycolpites sp.
Polycolporites sp.
Proteacidites auratus
Proteacidites sp.
Proteacidites thalmanni
Proteacidites tumidiporus ecollariatus
Pulcheripollenites inrasus
Pulcheripollenites krempii
Pulcheripollenites narcissus
Rhoipites bradleyi
Rhoipites pissinus
Rosannia manika
Scabrastephanocolpites albertensis
Scabrastephanocolpites lepidus
Senipites drumhellerensis
Singhia urwashii
Stelckia unica
Stelckia vera
Stelckia xenoforma
Symplocoipollenites vestibulum
Tetracolpites pulcher
Tricolpites parvistriatus
Trifossapollenites ellipticus
Tubulifloridites aedicula
Wilsonipites nevisensis
Wodehouseia gracile
Wodehouseia jacutense

Fig. 604. Fossil leaf.

Fig. 605. Fossil leaf.

Plant Parts of Unknown Affinities

Many unidentified plant parts are found in the formation representing possible dicotyledonous angiosperms. Although not formally identified, the preceding plant parts appear identifiable but have not been at the present time. Some such as the roots are preserved in multiple forms with or without cellular collapse.

LEAF REMAINS

Dicotyledonous angiosperm leaf floras are not commonly encountered in the formation but are present. None of the leaf floras have been studied to date and only a handful of individual leaf specimens have been studied (Bell 1949).

The leaf floras represent various environments. Some leaf beds, although containing many specimens, contain a much-reduced assemblage, with leaves from the immediate surrounding area of deposition. These remains are all considered to be from deciduous trees. Other sites appear mixed and, although transported, are not far removed from the source. The mixed flora can sometimes contain both angiosperm and conifer (i.e., *Parataxodium, Mesocyparis, Torreya, Cunninghamia*, etc.), but again the angiosperms appear from deciduous trees. Evergreen angiosperm forms are not yet identified in the flora. A variety of unstudied angiosperm leaves are shown (Figs. 604–612).

Fig. 606.   Fossil leaf.

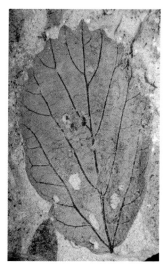

Fig. 607.   Fossil leaf.

Fig. 609.   Fossil leaf.

Fig. 610.   Fossil leaf.

Fig. 608.   Fossil leaf.

Fig. 611.   Fossil leaf.

Fig. 612.   Fossil leaf.

STEMS WITH CONVOLUTED INTERNAL STRUCTURE DUE TO TISSUE COLLAPSE

These stem sections (Figs. 613–616) may or may not belong to the same species. Externally they are unremarkable and appear as smooth woody sections up to 3.1 mm in diameter. Internally the tissues appear convoluted and collapsed. In some samples the epidermis is attached to the central core and in other samples it is free. It is figured here mainly due to the internal structure with the hope it may eventually be identified.

Fig. 613.   Fossil stem cross-section.

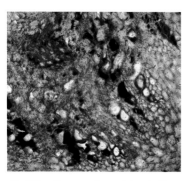

Fig. 614.   Close-up of vasculature in Fig. 613.

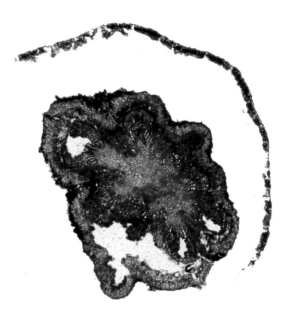

Fig. 615.   Close-up of vasculature in Fig. 616.

Fig. 616.   Fossil stem cross-section.

STEMS WITH SPOKED INTERNAL STRUCTURE

These stem sections (Figs. 617–622) appear very diagnostic but as yet are unidentified. Externally they are roughened by departing traces. Sections are up to 4.1 mm in diameter. Internally the morphology appears spoke-like due to the desiccation and collapse of the tissues. The departing vascular traces are prominent with vasculature present in both the inner and outer cortex. The central pith consists of larger-celled parenchyma, some containing dark contents. It is hoped that the unusual morphology of these stems internal structure may aid in its identification.

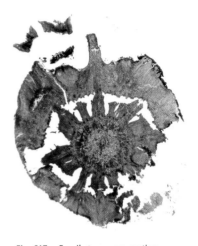

Fig. 617. Fossil stem cross-section.

Fig. 618. Fossil stem cross-section.

Fig. 619. Fossil stem tangential section of outer cortex showing a departing traces vascular organization.

Fig. 620. Fossil stem tangential section close-up showing spiral, bordered and scalariform pitting.

Fig. 621. Fossil stem tangential section showing pitting types.

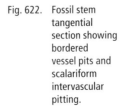

Fig. 622. Fossil stem tangential section showing bordered vessel pits and scalariform intervascular pitting.

Fig. 623.   Fossil stalk, basal view.

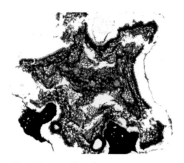

Fig. 624.   Fossil stalk, lateral view.

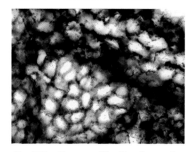

Fig. 625.   Fossil stalk cross-section.

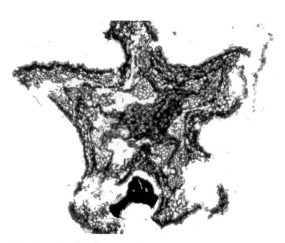

Fig. 626.   Fossil stalk cross-section.

Fig. 627.   Close-up of fossil stalk vascular
bundle.

Fig. 628.   Close-up of amber coloured cells in fossil
stalk.

FIVE-LOBED STALK

These plant stalks are unusual in that they contain well-formed abscission bases (Figs. 623, 624) and appear to have been five-lobed in life. The surrounding silica from which the tissue recessed appears lobed as well (Figs. 625, 626). The basal diameter is up to 3.2 mm. This plant had distinctive vascular morphology with cells containing amber-coloured contents (mucilage cells?) (Figs. 627, 628).

ROUND STEM WITH LATERAL VASCULAR

DEPARTURES

A very distinctive vascular pattern is exhibited in this stem and also shows how degradation due to cellular collapse affects some sections. Both are from the same piece, one at a node and the other at an inter-node. The stem is up to 1.9 mm in diameter (Figs. 629–631).

Another stem 2 mm in diameter (Figs. 632, 633) of similar morphology to the inter-node in Figure 631 was also found, but this may represent a different genus or family.

Fig. 629. Fossil stem showing cellular degradation.

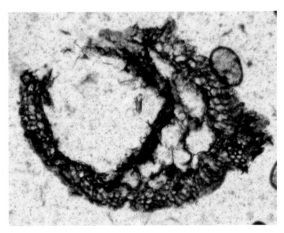

Fig. 630. Central core of Fig. 629.

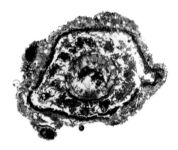

Fig. 631. Better preserved section of fossil stem.

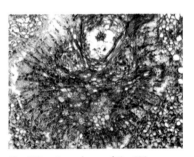

Fig. 633. Central core of Fig. 632. Compare to Fig. 631.

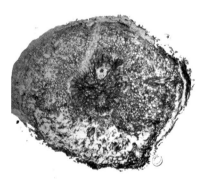

Fig. 632. Fossil stem with similar morphology.

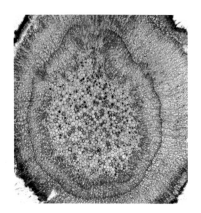

Fig. 634.　Fossil stem cross-section.

TRILACUNAR STEMS

Many angiosperms have wood of the trilacunar type. The specimens illustrated (Figs. 634–639) are from a fossil dicot. Stems are up to 3.3 mm wide and contain reduced interfascicular regions. It appears to be common at one site amongst woody debris. It is placed here mainly to show its presence.

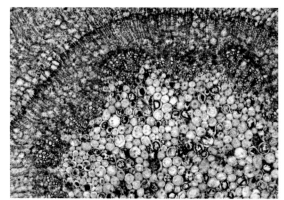

Fig. 635.　Fossil stem close-up.

Fig. 636.　Close-up of fossil stem departing traces.

Fig. 637.　Close-up of pith cells.

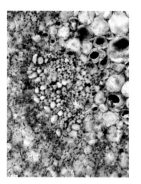

Fig. 638.　Close-up of median trace.

Fig. 639.　Close-up of lateral trace.

ROOTS

Roots are common in the formation. Many occur in the ironstones as silica replacements (Fig. 640). All higher plants produce them so a variety of roots are found. A few of the common ones are shown as well as one rare specimen.

These tetrarch roots (Figs. 641–647) are common in the formation at a variety of sites. They are unusual in that they appear preserved at many sites in differing ways. They can exhibit total vascular collapse, partial compression as if composed of soft tissue, and full preservation of all cell tissues. It is felt that they may have been herbaceous in nature. Roots are up to 3.2 mm in diameter and usually occur in masses.

This type of tetrarch root (Figs. 648–651) is also very common but has a wide range in size. A pentarch specimen (Figs. 652–654) is shown which may belong to a similar plant.

This last root (Fig. 655) is placed mainly to show the details that can be preserved externally. This specimen still retains its root caps (arrow).

Fig. 640. Ironstone containing root masses.

Fig. 641. Fossil root cross-section showing total vascular collapse.

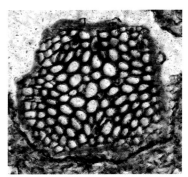

Fig. 642. Close-up of fossil root core from Fig. 641.

Fig. 643.   Dermal and epidermal cells showing thickenings. Close-up of Fig. 641.

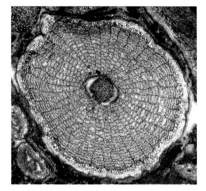

Fig. 644.   Well-preserved fossil root cross-section.

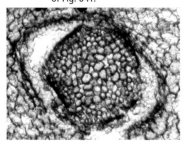

Fig. 645.   Tetrarch core of Fig. 644.

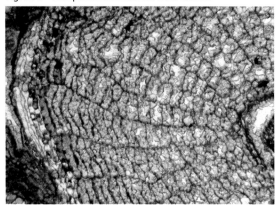

Fig. 646.   Close-up of cells from Fig. 644. Note thickenings on outer epidermal and dermal walls.

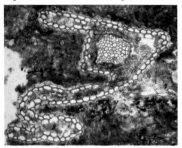

Fig. 647.   Similar fossil root showing tissue collapse.

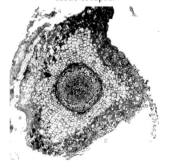

Fig. 648.   Fossil root cross-section.

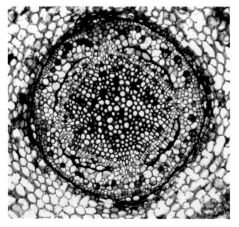

Fig. 649.   Close-up of Fig. 648 showing tetrarch vasculature.

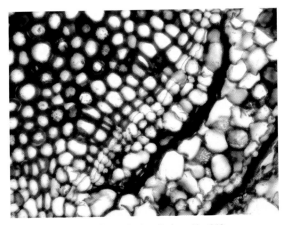

Fig. 650.   Close-up of vascular bundle from Fig. 648.

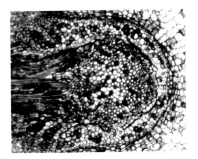

Fig. 651.   A fossil root tetrarch bundle with departing lateral root trace.

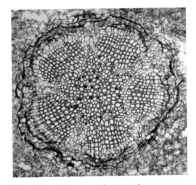

Fig. 653.   Close-up of pentarch vasculature from Fig. 652.

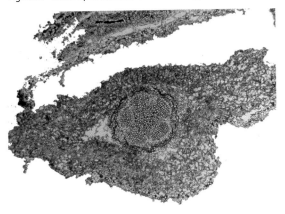

Fig. 652.   A pentarch fossil root cross-section.

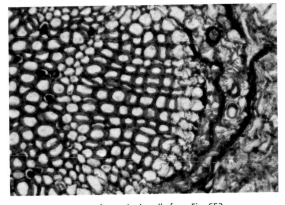

Fig. 654.   Close-up of vascular bundle from Fig. 652.

Fig. 655.   Silicified roots with intact root caps (arrow).

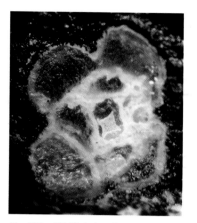

Fig. 656. Silicified plant fragment.

ROOT OR STEM?

This piece (Fig. 656) is unusual yet very diagnostic. It contains four large arenchymatous lobes with four smaller arenchyma areas juxtapositioned to them. The epidermis appears crenulated. The stele is preserved with a partially preserved endodermis. The specimen is small (3 mm in diameter) and is still in ironstone.

Seeds

Many different seeds of suspected dicotyledonous affinity can be found in the formation. Many of the smaller seeds are not shown due to the inability to recognize them in the field. They are important but appear mainly in acid preparations and would be too numerous to illustrate. Some of the more commonly collected or unusual ones are shown (Figs. 658–669)

ACER-LIKE SEEDS?

These seeds (Fig. 658) appear to have contained a raphe similar to a maple seed (Fig. 657) based on the remnants of the proximal raphe bar attachment (arrow). On closer examination of the fossil, there does not appear to be an attachment for a second seed. The similarity to *Acer* appears superficial. Pollen of the Aceraceae is also absent from the record.

Fig. 657. *Acer negundo* seed.

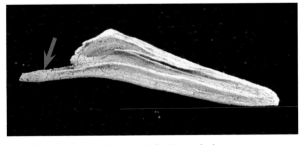

Fig. 658. Fossil seed with arrow indicating raphe bar.

Fig. 659. Ironstone nodule containing multiple seeds (red arrows).

LARGE OVAL SEEDS

These seeds are large, up to 1.2 cm long, woody, and contain two valves to the seed coat (Figs. 662, 663). They sometimes preserve light ridges running longitudinally with the axis (Fig. 660) or may be smooth due to erosion (Fig. 661).

Seeds are sometimes found as masses in ironstone (Fig. 659). They are common in some sites and may be angiosperm or possibly gymnosperm in origin. These seeds appear similar to those of seed type R (Serbet 1997).

Fig. 660. Fossil seed, dorsal view.

Fig. 661. Fossil seed, lateral view.

Fig. 662. Fossil seed lateral longitudinal section.

Fig. 663. Fossil seed cross-section.

FLAT-WINGED SEED

This seed (Fig. 664) was figured (Aulenback and Braman 1991) and described as seed type B (Serbet 1997). It is shown here as a complete specimen. The seed is bifacially flattened and up to 2 mm long. It has not been formally named but appears very distinctive.

CARPITES VERRUCOSA SEED

Carpites verrucosa (Figs. 665, 666) is a seed of unknown affinity. Previous debates (Lesquereux 1878; Krassilov 1976; McIver and Basinger 1993) placed them as leaves or seeds, but with the discovery of silicified forms from the Horseshoe Canyon Formation, it has become apparent that they are indeed seeds (McIver and Basinger 1993).

    C. verrucosa seeds are reniform in shape and up to 10 mm long with distinctive surface pitting. These seeds are sometimes confused with those of small leaves when found as compressions. As mixed leaves and seeds, these plant parts were previously placed as *Porosia verrucosa*. A seed collected from the Cretaceous outcrops of Grande Prairie, Alberta, shows the distinctive pitting on the surface of the carbon-trace form.

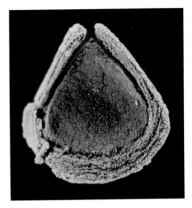

Fig. 664.   Fossil seed.

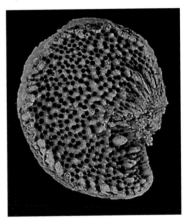

Fig. 665.   *Porosia verrucosa* silicified seed.

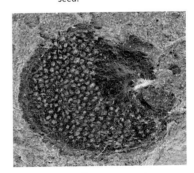

Fig. 666.   *Porosia verrucosa* carbon trace seed from the Campanian of Grande Prairie, Alberta.

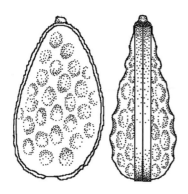

Fig. 667. Fossil elliptical seed, facial view left, lateral view right.

ELLIPTICAL PEBBLED SURFACE SEED

This seed is not photographed but a drawing is given (Fig. 667). The seeds are very small, up to 1.9 mm in length, and two-valved. The surface contains small divots on the faces and the lateral sides appear thickened. They appear diagnostic but are undescribed.

SCALLOPED SURFACE SEED

This 4.5 mm long seed is unusual as it is bi-lobed on the lower section (Figs. 668, 669). It too appears diagnostic but is undescribed.

Fig. 668. Fossil seed, lateral view.

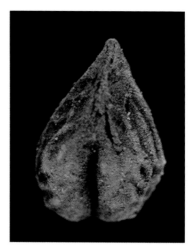

Fig. 669. Fossil seed, facial view of Fig. 668.

MONOCOTYLEDONAE (LILIOPSIDA)

The "Monocots," as the name implies, have a single (mono) seed leaf (cotyle), unlike the dicots that contain two seed leaves. Monocots are also distinguished by normally having floral parts in groups of three, although deviations occur (compare flowers in Figs. 670–679). Vascular bundles in their stems, which rarely have secondary growth, are usually scattered.

Fig. 670. *Amaryllis* sp. flower.

The monocots are presently the rarest order in the Cretaceous fossil record and have an enigmatic origin. Many have assumed the class branched off early from the dicots approximately 125 million years ago during Aptian/Albian times (Bremer 2000).

Although many aborescent (tree-like) monocot remains are found in the Upper Cretaceous of North America, the earliest monocots were thought to be of herbaceous nature.

Based on attributes of recent monocot growth habits, fossil monocots are thought less likely to fossilize. The majority of monocots are of herbaceous nature and have leaves that generally wither in situ and are not shed in large numbers (Herendeen and Crane 1995). Roots also wither in response to seasonal cues of wet/dry or hot/cold. Only the stem (rhizome) is likely to fossilize. Even so, living rhizomes do not transport well and break down quickly upon exposure due to their fleshy nature. There is also the difficulty in the field of recognizing poorly preserved monocot remains.

Early monocots are also thought to have been insect pollinated (zoophilous). Recent zoophilous plants generally produce small amounts of pollen thus fossilization of pollen is also less likely (Herendeen and Crane 1995).

Fig. 671. *Aponogeton distachyos*, water hawthorn flower.

Fig. 672. *Aspidistra elatior,* cast-iron plant flowers.

Fig. 673. *Sagittaria latifolia,* duck potato male flowers.

Fig. 674. *Tulipa* sp., tulip flower darwin hybrid group.

The earliest known monocotyledons appear in the Lower Cretaceous (Aptian) zone I of the Potomac Group (125 mybp). These are found as pollen of *Liliacidites* and plant leaf remains of *Acaciaephyllum*, although these finds are considered equivocal (Gandolfo et al. 2000).

The oldest known unequivocal monocotyledonous remains are from the Late Barremian to Early Aptian of Portugal (110–120 mybp). The fossil is a cutinized plant fragment covered in pollen. The fossil is placed in the Spathiphylleae; subfamily Monsteroideae (Friis et al. 2004).

In North America, monocotyledonous flowers are known from the Upper Cretaceous, Turonian Stage, Raritan Formation (90 mybp) of New Jersey. They are placed in the family Triuridaceae. It is felt that these particular plants were achlorophyllous (without chlorophyll), saprophytic and mycotrophic in habit (Gandolfo et al. 1998c; 2002).

In Alberta the oldest described monocot remains are those of *Zingiberopsis,* a ginger-like leaf from the Paskapoo Formation (Paleocene) (Hickey and Peterson 1978). Undescribed monocot rhizomes are also recognized from the Paleocene (Fig. 686).

Undescribed herbaceous rhizomes from the Dinosaur Park Formation (Cretaceous, Campanian) that are also ginger-like in form have been found (Figs. 680–683).

Even older yet are a variety of undescribed fossil herbaceous rhizomes with monocot attributes found in the Campanian, Oldman Formation (78.5–76.5 mybp), which undoubtedly represent a number of genera or families (Figs. 684, 685, 687, 744).

In the Horseshoe Canyon formation, a variety of undescribed fossil monocots are now known, mostly represented by stem and rhizome sections.

Fig. 675.   *Acorus calamus,* flowers.

Fig. 676.   *Sparganium simplex,* burr reed male flowers.

Fig. 677.   *Stromanthe sanguinea* flowers.

Fig. 679.   *Tricyrtis* sp. Toad lily flower.

Fig. 678.   *Chasmanthe aethiopica* flower.

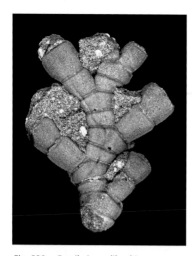

Fig. 680. Fossil ginger-like rhizome. Dinosaur Park Formation, Alberta.

Fig. 681. Ginger-like rhizome close up of surface cells. Dinosaur Park Formation, Alberta.

Fig. 683. Fossil corm close up of surface cells. Dinosaur Park Formation, Alberta.

Based on these plant remains, a minimum of seven to ten plant forms or types can be recognized from different preservational states (Figs. 720, 745–753)

Only three pollen types are recognized from the Horseshoe Canyon Formation. The paucity of pollen forms may again reflect our present inability to distinguish pollen forms due to non-diagnostic traits (see discussion on Taxodiaceae), or they may have been zoophilous, and the pollen may just be extremely rare and, if close to the source, not yet found.

Although large aborescent plants are present representing the Palmaceae, other orders such as the Pandales and the Scitamineae that were previously thought to have occurred in the formation are absent.

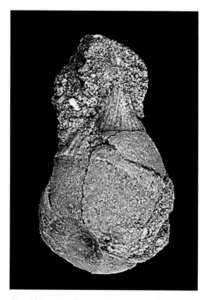

Fig. 682. Fossil corm. Dinosaur Park Formation, Alberta.

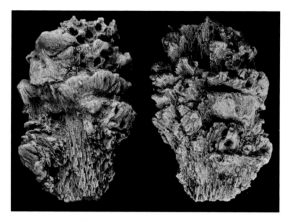

Fig. 684. Vertically orientated fossil rhizome, two views. Oldman
Formation, Alberta.

Fig. 685. Horizontally orientated fossil rhizome. Oldman
Formation, Alberta.

Fig. 686. Two connected fossil plants via rhizome and roots.
Paleocene, Alberta.

Fig. 687. Possible horizontal or vertically
orientated fossil rhizome.
Oldman Formation, Alberta.

Fig. 688. *Arisaema consanguineum,* flower.

Fig. 689. Fossil partial leaf with ovules attached. Oldman Formation, Alberta.

Alismatales

ARACEAE (ARUMS)

The family Araceae contains 106 genera and up to 4,025 species, which range from tree-like to herbs (Figs. 688, 696–700) and epiphytes.

The earliest occurrence of the Araceae is in the form of a cuticular fragment with pollen clumps assigned to the Spathiphylleae (subfamily Monsteroideae) form the Late Barremian to Early Aptian (110–120 mybp) of Portugal (Friis et al. 2004).

The oldest unofficial record from Alberta is from unpublished material from the Oldman Formation of southern Alberta, which shows primitive reproductive structures that appear Araceae-like. The immature forms of these structures consist of a leaf with multiple ovules attached to the surface (Fig. 689) similar to various genera in the *Araceae* (Fig. 698). Mature forms show a leaf that fuses along the front margin after fertilization (Figs. 690, 691). The leaf apex is modified into a cap-like structure that apparently opened to disperse the minute seeds. The plant is interpreted as an aquatic of ephemeral ponds where, during the dry season, the pods would open up, dispersing the seeds.

In the Campanian of the Horseshoe Canyon Formation, a highly diagnostic seed head named *Albertarum pueri* (Bogner et al. 2005) representing the Subfamily Orontioideae has been described. The fossil is up to 2.5 cm wide and represents a ¾ section of a seed head, which shows internal as well as external structure (Figs. 692–695). *Albertarum pueri* shares traits with the extant *Symplocarpus, Lysichiton,* and *Orontium,* all of which are in the subfamily Orontioideae. Although sharing traits of all three genera, the fossil is most similar to *Symplocarpus* (Bogner et al. 2005).

Symplocarpus contains only the single species *S. foetidus*, commonly called the Skunk Cabbage or Polecat Weed, which occurs in swamps and wet woodlands in northeastern North America and northeastern Asia. Pollen of the Orontioideae has not been found in the formation.

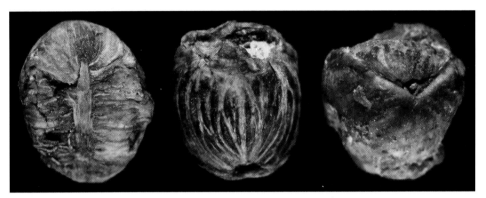

Fig. 690. Three views of fossil seed pods. Oldman Formation. Left, back view; middle, front view; right, oblique apical view (due to compression).

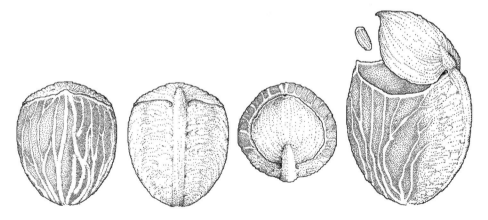

Fig. 691. Drawings of seed pod. Oldman Formation. From left to right: front view, back view, top view, back oblique view top open for seed dispersal.

Fig. 692. *Albertarum pueri*, external view.

Fig. 693. *Albertarum pueri*, basal view.

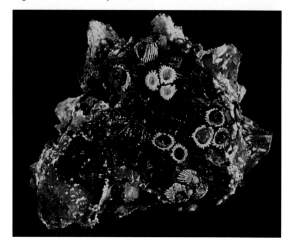

Fig. 694. *Albertarum pueri*. fractured back (internal) view.

Fig. 695. *Albertarum pueri*. view of exposed seeds.

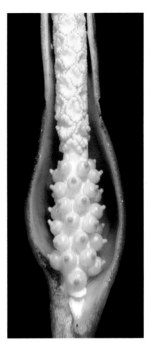

Fig. 696. *Arum italicum,* flower.

Fig. 697. *Peltandra virginica.* Dissected flower exhibiting ovules free of the spathe.

Fig. 698. *Pinellia ternata.* Dissected flower exhibiting ovules fused to the spathe.

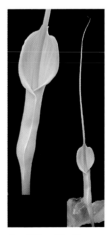

Fig. 700. *Pinellia ternata* close-up (left) and complete flower (right).

Fig. 699. *Arisaema triphyllum* flower.

Fig. 701.   Fossil seed bed with seeds in situ. Scollard Fm., Alberta.

Fig. 702.   Fossil seed. Scollard Fm., Alberta.

Arecales

ARECACEAE (PALMS)

The first palms known from Alberta are from the Foremost Formation (Late Campanian) with fronds of a *Sabal*-like form (compare Figs. 705 and 706 to Fig. 704).

In the Horseshoe Canyon Formation, the Palmaceae is represented by large, drupe-like seeds described under the fossil misnomers *Ficus ceratops* or *Guarea*, which have been described from other formations.

A massive and extensive bed of seeds and amorphous wood pieces is known (Fig. 701) from the overlying Scollard Formation. In the upper sections of the Horseshoe Canyon Formation, finds consist of rare individual seeds.

The seeds are large, up to 5 cm long and range in colour from grey or brown (Fig. 702) to black. The endocarp vasculature of the seed, when present, appears arilate-like (Fig. 703), although it would have been contained in a softer tissue during life.

These seeds are similar to those of *Spinifructus antiquus* from the Frenchman Formation (Cretaceous) of Saskatchewan (McIver 2002), although none of the Alberta specimens appear to bear the spines of the exocarp that are prominent in the Saskatchewan specimens. This may represent a preservational bias or possibly a species difference as many hundreds of specimens have been found in Alberta in situ without spines. *Spinifructus antiquus* seeds have been identified as belonging most likely to an extinct genus of Arecoid palm (McIver 2002).

A single rooting structure is known from the upper Horseshoe Canyon. The fossil consisted of a carbon-trace stem section with a large root mass. The fossil stem roots were restricted to one side with leaf scars on the opposite side. This implied a procumbent or trailing form.

Although frond material has not been found in the Horseshoe Canyon Formation, a section of frond was identified by the author in the field at the Scollard, Huxley T. Rex site in 1981. This site also produced seeds similar to those previously described.

Palms survived in Alberta until the earliest Paleocene based on fronds of a *Serenoa*-like representative recovered from Genesee (compare Figs. 708–707).

The living family consists of 187 genera with 2,000 species that range from warm temperate to tropical worldwide. The genus *Sabal* contains fourteen recent species. Sabals are evergreen and range in size from tree to shrub form. They have slow growth and prefer rich, moist well-drained soils in a protected partially shaded position. Warm temperate Sabals will tolerate light frosts. *Serenoa repens* (Fig. 707), the Saw Palmetto, from the southeastern United States is found in prairies, pine-lands, and sand dunes, and prefers moist rich soils in full sun. *Serenoa* is a spreading shrub that is drought- and frost-tender.

Fig. 703. Fossil seed in situ showing vascular markings. Basal view. Scollard Fm., Alberta.

Fig. 704. *Sabal chinensis* close-up of stipe apex.

Fig. 705. Close-up of stipe apex from Fig.
706. Foremost Fm., Alberta.

Fig. 707. *Serenoa repens* close-up of
stipe apex.

Fig. 706. Fossil palm frond. Foremost Fm., Alberta.

Fig. 708. Fossil palm frond. Paleocene, Alberta.

FOSSIL PLANTS

Liliales

LILIACEAE (LILIES)

The oldest known Liliidae is from the Upper Cretaceous, Turonian, Raritan Formation (90 mybp) of New Jersey (Gandolfo et al. 2002).

As was mentioned, many herbaceous monocot rhizomes representing different genera are known from the Horseshoe Canyon Formation. Undoubtedly some represent the Liliidae, but unfortunately diagnostics of the rhizomes in the many monocot families have not been studied in detail so identification of fossil families would be difficult at best.

Fig. 709. *Hemerocallis* sp. day lily.

Seed heads are extremely rare with only one found that possibly falls in the family (Figs. 711, 712). The fossil is 9 mm long and contains locules in sets of three with one ovule per ovary based on cuticle found in the ovary. This is a terminal compound structure with three separate flowers represented. This structure appears very advanced in construction with its fused floral organization.

The Liliaceae is a large family with up to sixteen genera and 635 species.

Fossil pollen:
Liliacidites mirus
Liliacidites morrinensis
Liliacidites variegatus

Fig. 710. *Liliacidites* sp.

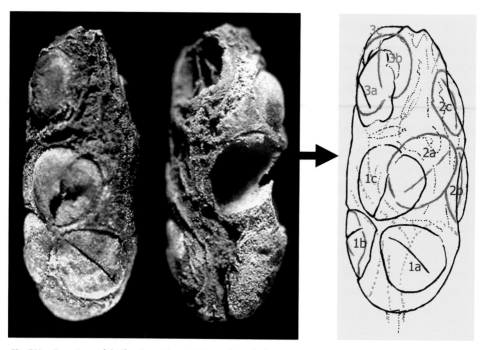

Fig. 711. Two views of the fossil seed head with see-through drawing showing orientation of the three flowers.

Fig. 712. Fossil seed head.

Poales

CYPERACEAE (SEDGES)

The Poales consists of seventeen families, 997 genera with up to 18,325 species.

The fossil record of the Poales is based on silicified seeds and a single stem section.

Cyperus-like

Cyperus-like seeds up to 1.5 mm long, preserved as calcium carbonate, have been found in the Campanian, Oldman Formation of Alberta (Fig. 713).

In the Horseshoe Canyon Formation, silicified seeds have been recovered. These seeds are up to 2.5 mm long, triangular in shape and drupe-like (Fig. 714).

A stem section (Figs. 715–719) has also been identified as Poales, most possibly in the Juncaceae. The stem appeared non-descript based on external morphology, but its internal anatomy was mainly intact. The stem is up to 2.1 mm wide and consists of collateral vascular bundles restricted to a peripheral zone.

The Poales is cosmopolitan in distribution.

Fig. 713. Three views of fossil seeds from the Campanian, Oldman Formation, Alberta. Top to bottom: apical, lateral, and basal views.

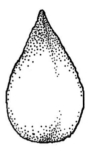

Fig. 714. Drawing of fossil seed figured by Serbet (1997). Drawn with permission.

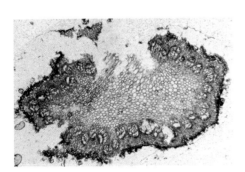

Fig. 715. Fossil stem cross-section.

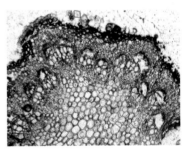

Fig. 716. Close-up of right portion of Fig. 715.

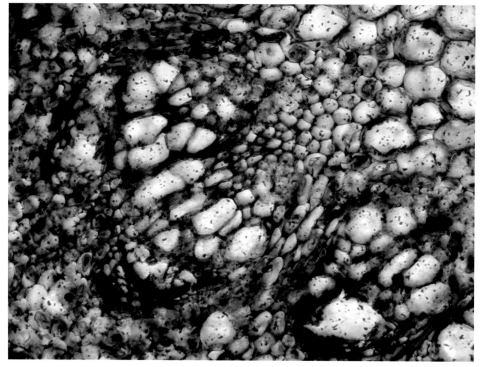

Fig. 717. Close-up of collateral vascular bundles. Note the two fused bundles (middle and upper left).

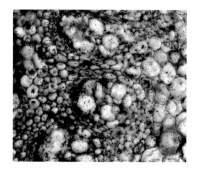

Fig. 718. Close-up of individual collateral vascular bundle.

Fig. 719. Close-up of central parenchyma.

Incertae Sedis

Monocot Rhizomes and Plant Parts of Unknown Affinity

TRI-LOBED RHIZOMES

A variety of rhizomes have been found that all follow the same basic internal and external form, although growth patterns vary. The rhizomes range from globose to linear and are up to 1.5 cm in length (Figs. 720–728). The rhizomes contain root, inflorescence scars, and lateral rhizome connectives mixed in three distinct, equally spaced vertical rows. Leaf vascular strands are distinct in single horizontal rows between root and inflorescence vertical rows. Roots scars are circular with a central vasculature present and commonly protruding.

Inflorescence scars are round with a smooth narrow band of outer basal tissue (Fig. 727). Cortex tissue of these scars is raised but flat-topped with underlying vasculature, giving the surface a pebbled appearance. Based on the smoothness and consistent level of the scars, the inflorescence is interpreted as terminal and deciduous. Inflorescences occur both laterally and apically.

Lateral rhizome connectives to other rhizomes are two or three times larger than the root scars (Figs. 720–722, 725) and occur laterally and basally. Root scars form a ragged ring in ring morphology with the central ring in-filled by vasculature. Lateral rhizome connectives are sometimes found with similar morphology.

External leaf vasculature is in horizontal rows with individual leaf vascular strands in a single distinct linear row, either preserved as indents due

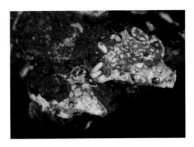

Fig. 720.   Original surface find of fossil colonial rhizome.

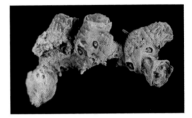

Fig. 721.   Fossil colonial rhizome.

Fig. 722.   Fossil colonial rhizome with large and small rhizomes.

Fig. 723.   Fossil rhizome.

Fig. 724.   Fossil rhizome.

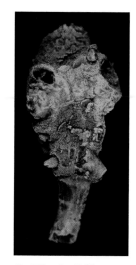

Fig. 725.   Fossil rhizome
with basal
connective.

Fig. 726.   Fossil rhizome with apical
connective.

Fig. 727.   Fossil rhizome lacking
apex.

Fig. 728.   Spiraling fossil
rhizome.

FOSSIL PLANTS

to decay or as partially eroded strands. From six to nine vascular strands occur per leaf row (Figs. 721–728).

Internal anatomy shows a thin epidermis of one or two cells. An endodermis is present, delimitating a wide cortex and central cylinder (Figs. 729, 730). The cortex contains concentric more or less uniformly distributed amphivasal bundles (Figs. 729–732, 734–738, 740–742). There are no large differentiated metaxylem elements (Figs. 735–738).

Smaller bundles are near the periphery (Figs. 729, 730, 732). The central cylinder lacks vasculature and consists of parenchyma cells (Figs. 731, 739) and some rare cells containing dark contents.

Vascular bundles for individual vegetative traces are linear and straight, departing from in the cortex at the same level, forming a distinctive band (Figs. 729, 733).

Root traces depart individually at the boundary of the cortex and epidermis (Fig. 729).

Lateral rhizome connectives originate in the cortex from multiple vascular strands at various levels and distances (Figs. 729, 730, 732).

It is hoped with this detailed description someone may eventually be able to place these rhizomes in a family.

Calcite-replaced rhizomes with identical internal and external morphology have been found in the Oldman Formation of southern Alberta (Fig. 744). It appears that the plant rhizome form survived for a few million years without change in the province.

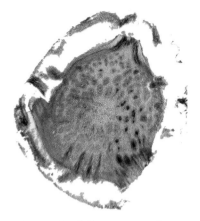

Fig. 729. Fossil rhizome cross-section showing morphology.

Fig. 730. Fossil rhizome cross-section showing morphology.

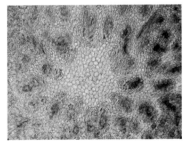

Fig. 731. Close-up of central core containing parenchyma cells from Fig. 730.

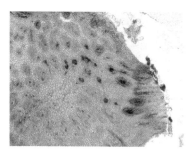

Fig. 732. Close-up of departing lateral connective vasculature from Fig. 729.

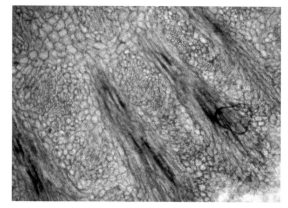

Fig. 733. Close-up of departing leaf traces from Fig. 729.

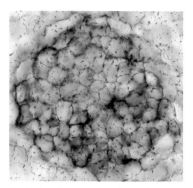

Fig. 734. Close-up of amphivasal vascular strand.

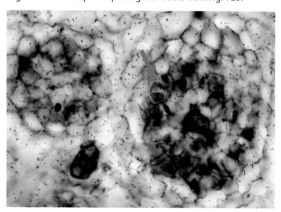

Fig. 736. Close-up of one maturing and one immature amphivasal vascular strand. Note dark contents in cells (arrow).

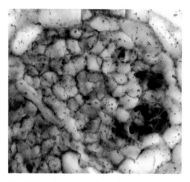

Fig. 735. Close-up of amphivasal vascular strand.

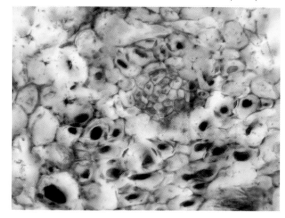

Fig. 737. Close-up of mature amphivasal vascular strand.

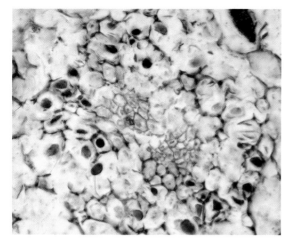

Fig. 738. Close-up of two fused mature vascular strands. Lower right appearing collateral with xylem and phloem internal.

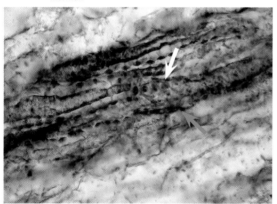

Fig. 741. Close-up of individual strand showing bordered vessel pits (white arrow) and tracheid with wide spiral thickenings (green arrow).

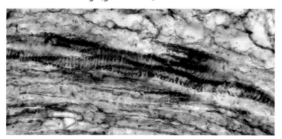

Fig. 742. Close-up of vascular strand showing scalariform intervascular pitting.

Fig. 739. Close-up of central cylinder parenchyma cells.

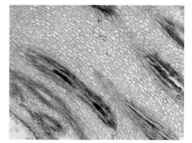

Fig. 740. Close-up of tangential section showing vasculature.

Fig. 743. Close-up of strands showing both alternate intervascular pitted vessels (green arrows) and scalariform intervascular pitting (brown arrow).

Fig. 744. Fossil rhizome from the Oldman
Formation, Alberta.

Fig. 745. Fossil rhizome.

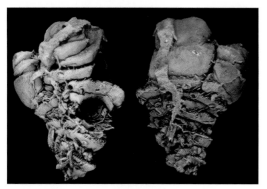

Fig. 746. Two views of fossil rhizome.

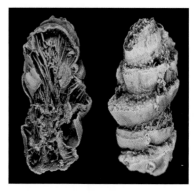

Fig. 747. Two views of fossil rhizome.

Fig. 748. Two views of fossil rhizome.

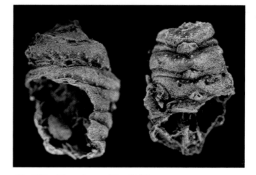

Fig. 749. Two views of fossil rhizome.

VERTICALLY ORIENTATED RHIZOMES

A variety of vertically orientated rhizomes have been found. These most likely represent more than one genus as they are from various localities in the formation and have slight differences, which may be of taxonomic importance. Not all are etched from rocks. Figure 743 is a large surface collected specimen 4.3 cm long. All the others (Figs. 746–749) are under 3 cm in length. Unfortunately, none have been studied.

SEMI-PROCUMBENT RHIZOME

This rhizome (Figs. 750, 751) found at the Drumheller Golf Course appears very different from the erect rhizomes. The rhizome consists of three growth heads, two of which appear pre-depositionally abortive. The rhizome is 5 cm long with massive leaf scar vasculature made up of possibly fifty to a hundred vascular strands per leaf base. The rhizome is robust with tight leaf scar internodes and roots randomly disposed. It remains unstudied.

LATERALLY ORIENTATED RHIZOMES

A few rhizomes have been found, which indicate that they were most likely laterally orientated in the soil during life. They are shown here mainly to indicate the variety of forms possible in rhizomes as a whole.

The first rhizome (Fig. 752) shows well-defined nodes and is 8 mm long. The second rhizome, which may represent a dicot (Fig. 753), contains large lateral buds on a thin rhizome. The total length is 9 mm.

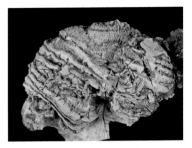

Fig. 750.   Fossil rhizome lateral view.

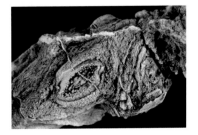

Fig. 751.   Fossil rhizome apical view.

Fig. 752.   Fossil rhizome.

Fig. 753.   Fossil rhizome.

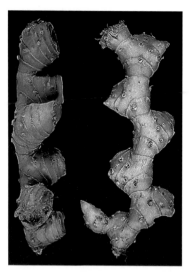

Fig. 754. *Costus speciosus*, ventral (right) and lateral (left) view.

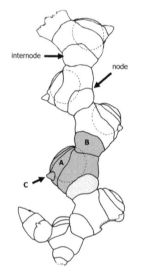

internode ➤
node

B
A
C ➤

Fig. 755. Drawing of *Costus speciosus* from Fig. 754, illustrating node and internode and the three main sections of a rhizome. A = main axis, B = secondary axis, C = tertiary axis.

Monocotyledon Rhizome Morphology: What's the Difference?

What is a rhizome? A rhizome is basically an underground stem. In angiosperms rhizomes occur in both dicots and monocots. A rhizome can be a corm or tuber but is not a bulb. A corm acts as a vertically positioned rhizome with reduced internode areas. A tuber, such as the potato, is attached to the parent plant, contains dormant buds, and is capable of vegetative reproduction. A bulb consists of many modified swollen leaf bases attached to a highly reduced stem, with basal roots such as the onion.

The study of monocots has focused mainly on the morphology and evolution of the above-ground vegetative and floral organs with little work on the rhizome. Many extant descriptions contain only the words "rhizome" or "corm" when referring to the mode of growth, and there are few detailed studies of rhizome morphology in extant genera.

The study of rhizome morphology in all monocot genera is a daunting task but much needed if we are to understand the evolution of the monocots in the fossil record. It is slowly becoming apparent that rhizomes are the most likely plant part to fossilize.

In linear forms, rhizomes of a single species may vary in shape and size due to nutrient availability, growth medium composition, or crowding, but the underlying morphology remains intact. This variability is evident in the fossil tri-lobed rhizomes (Figs. 720–743).

Root-production areas, leaf-scar morphology, lateral vegetative shoot placement and form, leaf internode vascular construction, external surface cell shape and form, internal vascular cylinder construction, vascular strand construction and ontogeny are constants that can be used in unison for rhizome description and identification.

The majority of linear procumbent monocot rhizomes are based on a repetitive plan. All consist of nodes, internodes, roots, buds, and an apex. Accessory organs such as lateral rhizome connectives and corms may also be present.

Internodes are stem-growth areas before and after the production of leaves, while nodes are where leafy vegetation is or was produced. The node is the scar left by a leaf. It may be pronounced or much reduced.

Roots are produced only in the internode. They may be restricted to one side of the rhizome (Fig. 758) or be found on all sides (Fig. 769). They can be produced in linear form (Fig. 773) or scattered (Fig. 766).

Lateral buds are usually produced in conjunction with the internode (leaf scar) immediately after it and with only one bud produced per internode (Figs. 754–757, 773). Buds can be alternate or helically arranged up the rhizome. This is similar to the production of a bud in the axil of a leaf base on a branch. Lateral buds can be arrested or active. Active buds form the extension of the rhizome (Figs. 754, 767, 773). Arrested buds allow for elongation of the rhizome or growing tip by the addition of multiple nodes prior to lateral development of an active bud or foliage production at the apex.

Unlike a tree branch, a rhizome is more a collective of individual plants, forming a colony of attached clones. The clone consists of three parts or axes, which together can be termed a *foliar unit*.

The primary axis is interpreted as producing the floral foliage and is terminal (Figs. 755A, 757A). The secondary axis continues as the non-foliage-

Fig. 756.   Fossil rhizome from the Dinosaur Park Formation.

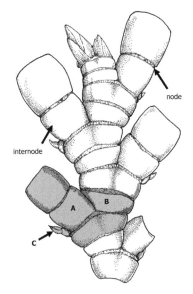

Fig. 757.   Drawing of fossil rhizome from Fig. 756.

bearing main rhizome body (Figs. 755B, 757B). The tertiary axis contains the laterally arrested bud (Figs. 755C, 757C). All axes may contain varying numbers of node/internodes.

This is the basic morphology of rhizomes, corms, and tubers. In this basic morphology, an amazing array of forms can be produced in a genus or species. There is much to learn from these plant parts (see Figs. 758–773).

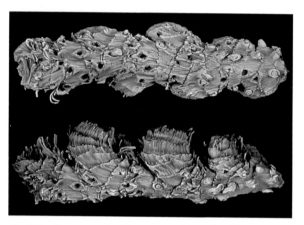

Fig. 758.   *Cyperus ustulus* rhizome, ventral (upper) and lateral (lower) view.

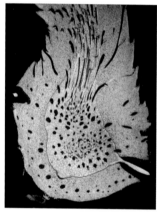

Fig. 759.   *Cyperus ustulus* rhizome, cross-section through diverging A and B axes.

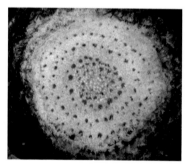

Fig. 760.   *Cyperus ustulus* basal leaf vasculature cross-section.

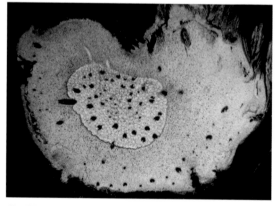

Fig. 761.   *Cyperus ustulus* rhizome inter-nodal cross-section.

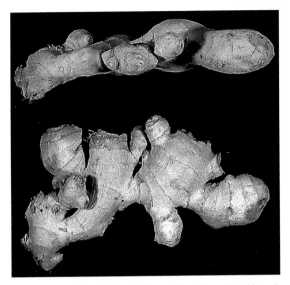

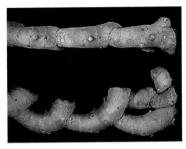

Fig. 763. *Hydichium coronatum* rhizome, ventral (upper) and lateral (lower) view.

Fig. 762. *Zingiber officianale* rhizome, dorsal (upper) and lateral (lower) view.

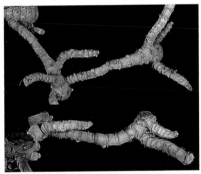

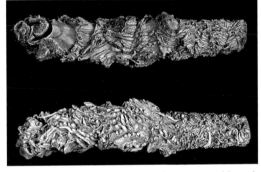

Fig. 764. *Alpinia zerumbet*, rhizome, ventral (upper) and lateral (lower) view.

Fig. 765. *Pontederia montevidensis* rhizome, ventral (upper) and dorsal (lower) view.

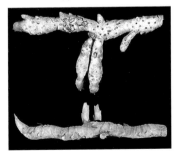

Fig. 766. *Iris* sp. rhizome, ventral (upper) and lateral (lower) view.

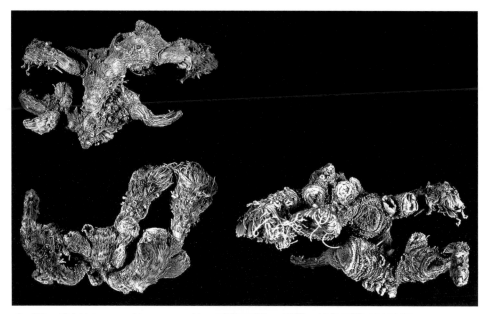

Fig. 767. *Alpinia purpurata* rhizome, ventral (upper), lateral (lower left), and dorsal (lower right) view.

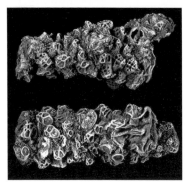

Fig. 768. *Nymphae* sp. rhizome, two views.

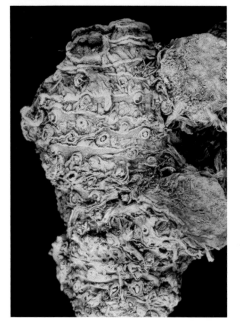

Fig. 769. *Spathyphyllum wallisii* rhizome, lateral view.

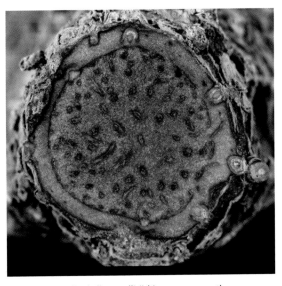

Fig. 770. *Spathyphyllum wallisii* rhizome cross-section.

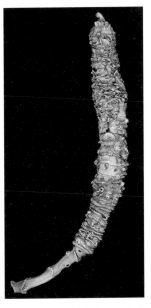

Fig. 773. *Acorus calamus* rhizome, ventral view showing zigzag root rows.

Fig. 772. *Spathyphyllum wallisii* rhizome lateral view showing basal connective.

Fig. 771. *Spathyphyllum wallisii* rhizome, lateral view.

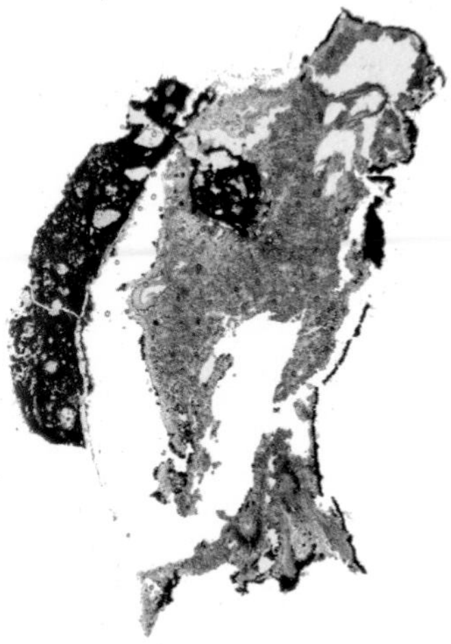

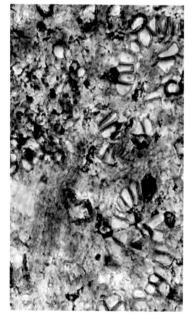

Fig. 774. Fossil rhizome cross-section.

Monocots Identified by Internal Structure Only

There are many amorphous plant remains that are non-descript externally. These may have a smooth surface or be eroded and pitted. It is only upon sectioning that these plant parts yield information on their affinities. The majority are dicotyledonous remains, but some rare ones are monocotyledonous.

Monocot 1

This specimen is 1.0 cm in diameter. The specimen has undergone desiccation and dissolution of cells in the cortex. The central cylinder, which appears to have been well defined, contained the leaf traces. The leaf traces are for the most part degraded (Fig. 774).

The cortex consists of irregularly formed amphivasal axial bundles of usually circumferentially discontinuous tracheary elements (Figs. 775, 776). These bundles are without a fibrous core or protoxylem.

The morphology of this rhizome suggests the Alismatales and possibly the Araceae. The specimen has not been studied.

Fig. 776. Fossil rhizome amphivasal bundle close-up showing tracheary element with scalariform intervascular pitting (arrow).

Fig. 775. Fossil rhizome close up of tracheary elements.

Monocot 2

This is a small specimen up to 5 mm in diameter (Fig. 777). It contains multiple vascular strands with varying degrees of preservation (Figs. 778–780). There is no central cylinder and the vascular strands are evenly spaced throughout the cortex. The ground tissues have suffered collapse.

The bundles are collateral with a surrounding bundle sheath. There are no large differentiated metaxylem elements.

Although a monocot, its affinities are unknown at present.

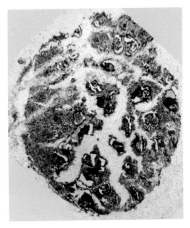

Fig. 777.   Fossil rhizome cross-section.

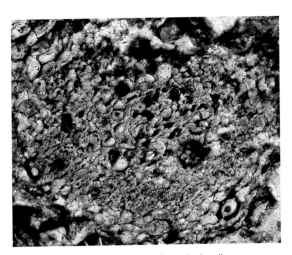

Fig. 778.   Fossil rhizome close-up of vascular bundle.

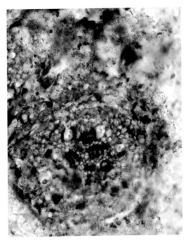

Fig. 779.   Fossil rhizome close-up of vascular bundle.

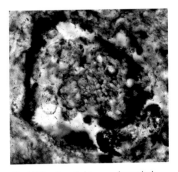

Fig. 780.   Fossil rhizome degraded vascular bundle.

Fig. 781. Fossil rhizome cross-section.

Fig. 782. Fossil rhizome cross-section showing leaf vasculature (arrow).

Monocot 3

This is a somewhat large specimen up to 1.3 cm in diameter. The specimen has undergone desiccation, degradation, and tearing of internal tissues, but the overall morphology of the vasculature appears intact.

Well-formed vascular strands interpreted as leaf traces are few in number and are restricted to a central cylinder (Figs. 781, 782), which is defined by its physical position but not bound. The leaf traces range from amphivasal to collateral. Many contain an interrupted or semicircular external band of tracheary elements around the core elements (Figs. 786, 789). Over the phloem can sometimes be found a weak sheath which also surrounds the tracheary elements (Figs. 784–786, 788). The xylem bundle contains non-conspicuous metaxylem elements (Figs. 784–789).

Axial bundles are present but consist of discontinuous linear tracheary elements or clumps dispersed throughout the cortex (Fig. 783). The majority of fundamental tissue has undergone major collapse prior to fossilization.

The epidermis appears to have been thickened and contained scattered cells with dark contents. These cells are interpreted as tannin cells (Fig. 791).

A departed strand, which is interpreted as a leaf vascular departure, is also present (Figs. 782 arrow, 790).

This fossil is interpreted as an erect underground rhizome of a small herbaceous monocot. It appears to represent a possible Alismatales similar to monocot 1. This specimen has not been studied.

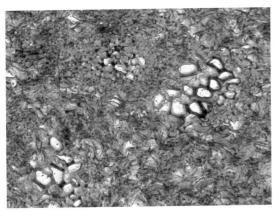

Fig. 783. Collapsed cortex showing dispersal of tracheary elements from Fig. 782.

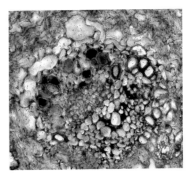

Fig. 784. Close-up of a collateral vascular bundle (Fig. 782). Note the occurrence of sporadic tracheary elements around the xylem and phloem.

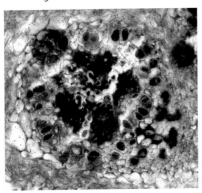

Fig. 785. Close-up of amphivasal vascular bundle (Fig. 782).

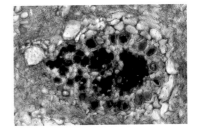

Fig. 786. Close-up of amphivasal vascular bundle (Fig. 782).

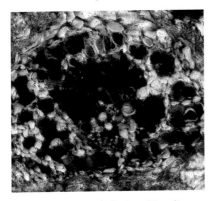

Fig. 787. Close-up of mixed amphivasal/ collateral vascular bundle (Fig. 782).

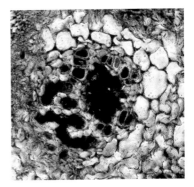

Fig. 788. Close-up of amphivasal vascular bundle (Fig. 781).

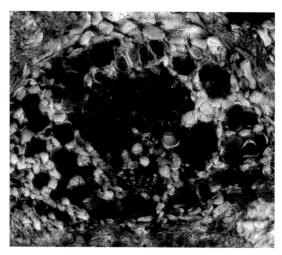

Fig. 789.   Close-up of amphivasal /collateral vascular bundle
(Fig. 781).

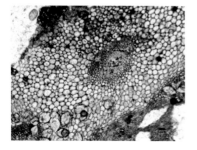

Fig. 790.   Vascular strand from leaf (Fig.
782).

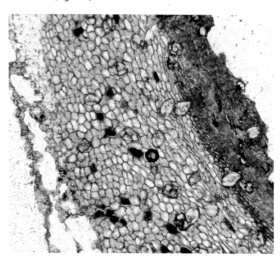

Fig. 791.   Close-up of epidermal layer (Fig. 782).

Possible Monocot Remains

Immature Seed Head

This seed head (Fig. 792) was found in an ironstone mass, which produced some *Trapago* leaf fragments and the unidentified Azollaceae. It appears to have been from an aquatic plant growing in the area. The seed head is 3 mm long and has multiple immature ovules held on an angulated base that appears to have consisted of soft tissues.

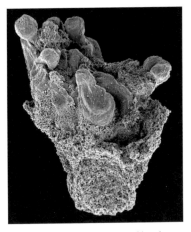

Fig. 792.   Fossil immature seed head.

Bi-valvate Seed Pod

These seed pods (Fig. 793) are up to 6 mm long and appear to have consisted of two valves that faced each other. Based on the flattened lower portion of the individual pods, it appears that two pods were attached at the lower half with the upper portion free. Remnants of the stalk position can also be seen.

Seeds are up to 3 mm long, have an open apex, and are irregularly wedge-shape, based on their position in the pod. The seed apex, although open, does not appear abraded as all seeds of this type, whether dispersed or in pods, have this morphology.

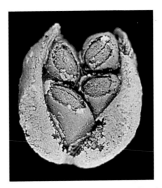

Fig. 793.   Fossil bi-valvate seed pod.

Individual seeds appear similar to those of *Disanthus hercynicus* from the Upper Cretaceous of central Europe (Knobloch and Mai 1983). Although individual seeds from Europe are placed in *Disanthus* of the Hamamelidaceae, the pod appears unlike *Disanthus*. The pod morphology is similar to that found in herbaceous monocots and is included here for the time being.

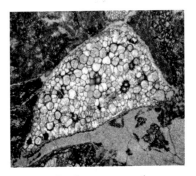

Fig. 794.   Fossil stem cross-section.

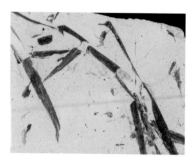

Fig. 795.　Fossil monocot leaves.

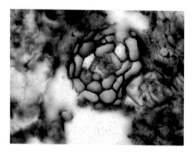

Fig. 796.　Fossil monocot root stele?

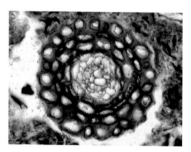

Fig. 797.　Fossil monocot root stele?

Stem Cross-sections

Many small stem fragments are encountered in the thin sectioning of sediments. This one appears to be of possible monocot origin (Fig. 794).

Linear Strap-shaped Leaves

These linear monocot leaves (Fig. 795) occur in a volcanic ash that sits on a coaly mud. The plants are rooted in the mud, but unfortunately techniques at this time do not allow the viewing of the rhizome. If worked in the future, this site may eventually produce reproductive structures as well as rhizomes.

Monocot-like Roots?

These roots (Figs. 796, 797) represent remains of stele sections with a totally collapsed epidermal and dermal region. This form of stele is found in various monocot groups but may also occur in certain dicots. They are included here to show their presence.

Storage Organs?

These specimens range from 1 to 2 cm long and are preserved as calcite (Fig. 798) and silica in the formation. They do not appear to be leaves as they show attachment at both ends and contain a thickened body.

They are similar to storage organs (tubers) but are dorso-ventrally compressed (3 mm thickness). Large vascular strands run the length of the organ on the outer surface. In silicified material, they contain an empty flattened hollow central core.

The fossil preservation in silica as dorso-ventrally compressed organs is unusual as silicified material usually retains its original shape. As a storage organ one would have expected an inflated

spherical shape. These may possibly be an underwater float organ due to the hollow centre noted in
silicified specimens, but again the organs are not
round. Presently they remain unidentified.

Seed Cuticles

Seed cuticles commonly occur during the preparation of palynological samples. A seed cuticle is
a resistant plant part that consists of cutin. They
may represent either inner or outer seed cuticle
layers, of which there may be many. Many cuticle
remains from the formation are known but undescribed, save the one below.

Cuticle remain: *Costatheca* sp.

Unknown

Many other unidentified plant parts occur in the
formation but are too numerous, or non-descript,
to illustrate. One that is unusual in its construction is seen in figures 799–801. These may prove
to not even be of plant origin. Originally thought
to be gastropod radula, this was discounted due to
the variability of size and form.

The forms were extracted from ironstones
along with a variety of plants (fungi sphericals, intact Azollaceae plants, various monocots, *Equisetum* and *Parataxodium*), and animals (gastropods,
Caddis fly puparia).

The "plant" form, preserved as silica, seems to
become more complex with size. Specimens range
in size from 5 to 15 mm in length. The surface appears to have been spiculate in specific areas with
the spicules pointed towards the main body (Fig.
801). All forms appear to be complete in length
without attachment points to other objects. They
are a true enigma at present.

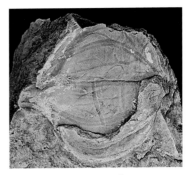

Fig. 798.   Calcite preserved storage organ.

Fig. 799.   Small fossil unknown spiculate
mass.

Fig. 800.   Large fossil complex unknown
spiculate mass.

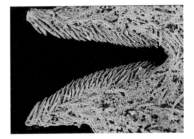

Fig. 801.   Close-up of intact spiculate area.

ENVIRONMENTAL RECONSTRUCTION OF THE HORSESHOE CANYON FORMATION BASED ON THE FLORA

It is well known that during the Upper Cretaceous overall temperatures were dropping within North America and reached a maximum low during the Maastrichtian (Upchurch and Wolfe 1993). Does the flora from the Horseshoe Canyon Formation reflect this cooling trend? Was the Horseshoe Canyon Formation tropical, subtropical, or warm temperate?

Past interpretations of the Horseshoe Canyon environment were based on the angiosperm palynomorphs. Based on these palynomorphs, it was concluded that the climate was subtropical to tropical as most of the palynomorphs had more plant genera or species in subtropical to tropical zones (Srivastava 1970).

Although pollen and spore types are referable to recent families and genera and are useful at interpreting environments, many palynomorphs do not show generic or specific affiliation in the Horseshoe Canyon Formation. Because of this, palynomorphs alone are not accurate for interpreting the environment in the formation. In any given family or genus, species may exist in many climatic zones, i.e., the genus *Pinus* ranges from subarctic to subtropical. Palynomorphs, however, are extremely important at showing floral composition and evolution.

In the past, in the Horseshoe Canyon Formation, it was concluded that the minor warm temperate elements (pollen) were most likely blown in from distant areas (Srivastava 1970). Unfortunately all the pollen that could be ascribed to living genera in subtropical and tropical zones do contain plant species that live in warm temperate environments as well. Few of the conifer or Cryptogam (ferns and fern allies) palynomorphs were used nor were any plant mega-fossils included. The present interpretation has attempted to use many of the palynomorphs and plants combined.

It can be seen that as one moves through the valley from the lower to upper Horseshoe Canyon that the environments change. The lower Horseshoe Canyon contains an abundance of coal zones that become more distantly spaced in the upper Horseshoe Canyon. Coals are indicative of wet environments. The presence of marine sediments indicates the proximity of the Bearpaw Sea (Western Interior Seaway) in the lower portion of the formation and validates this interpretation.

Coal zones and marine incursions would seem to suggest moist to wet coastal environments, while the distance between coal zones in the upper sections of the formation suggests a drying trend, and more fluvial environment.

Floristic analysis using an evaluation of the floral composition may give the best guess of temperature and moisture in the Horseshoe Canyon palaeoenvironment. The nearest living relative method, although most useful, must be used with some reservation. Problems arise when the related recent genera are restricted to refugia.

KEVIN R. AULENBACK

Recent genera may be occupying climatic niches that are not optimum for the species. These stressed refugia may have no outlet for dispersal (other than human-induced). *Sequoiadendron*, *Metasequoia*, and *Ginkgo* are prime examples. These genera show great ability to re-colonize palaeo-habitats once occupied (i.e., southern United States, central Europe) when reintroduced. Even genera such as *Glyptostrobus*, which is thought to be extinct in nature (post-extinction refugia), may be a poor climatic or floristic indicator.

With this in mind, the nearest living relative method relies on fossil taxa related to a modern genus and the nearest living relative being identified at the species level. The mean annual temperature and mean annual range in temperature for the flora is estimated based on the habitat preferences of the living relatives, and from this information a thermal regime of the palaeoenvironment is determined (Dolph and Dilcher 1979).

Of the fifty-three families represented by both fossil plant remains and palynomorphs only eighteen can be identified to genus with five genera being wholly extinct. Of the remaining, three belong in the cryptogams, seven belonging to the conifers and three in the angiosperms.

The choice of a nearest living species for these genera is easiest for the *Phylloglossum*, Equisetales, some of the Filicopsida, and conifer element due to the reduced number of species in each genus identified. The angiosperms chosen may or may not represent the most logical species based on the large number of species in each genus as well as incomplete study of the fossil specimens at hand.

The *Phylloglossum*, which is only represented by palynomorphs, is presently monotypic with *P. drummondii*, which is restricted to eastern Australia. It prefers moist wet habitats.

In the Equisetales two fossil forms are known. These forms appear most relatable to the two extant species, *E. sylvaticum* and *E. arvense*, which bear tubers as well. Both extant species presently inhabit moist to wet habitats with an almost worldwide distribution.

In the Filicopsida, the fossil Azollaceae, although representing an early evolutionary line, is interpreted to be a direct ancestor of modern *Azolla*. As such, it is felt that its modern equivalent would be *A. pinnata*, which is thought to be the most primitive of the living forms. *A. pinnata* inhabits slow to still waters in many differing habitats in Africa, India, Asia, and Australia.

The *Azolla conspicua* spore form is considered an extinct line without lineage to any extant Azollaceae.

In the Salviniaceae, there is the newly found extinct fossil (Figs. 132–147), which hints at a possible ancestor for the family. This fossil, however, does not produce any evolutionary forms in the family to the present.

Dorfiella (including *"Hydropteris"*) does appear to be the evolutionary ancestor for *Salvinia*, but a species relatable form is not identifiable due to *Dorfiella*'s primitive form being far removed from the living forms.

In the Marsileaceae, there has not been conclusive proof of the existence of any of the three living genera in the Cretaceous. The spore genus *Molaspora* is unlike the living genera and presently stands alone as a precursor to modern forms. Although felt to represent a proto-*Marsilea* based on evolutionary interpretations given by other authors, this standing is equivocal without more complete remains besides sporocarps and suspected foliage.

Even if representing a proto-*Marsilea*, the large number of extant *Marsilea* species and the incomplete remains of the fossil sporocarp make it unwise to pick an extant species representative.

The fossil *Osmunda* has been related to the extant *Osmunda cinnamomea*. Although the fossil species placement may be questioned here, the fossil still represents an *Osmunda* species closely related to *O. cinnamomea*. *O. cinnamomea* inhabits moist wet habitats in North America, South America, the West Indies, and east Asia. In North America, it grows in the vicinity of *Taxodium* and *Torreya* and is considered a warm temperate to temperate species.

The *Anemia* would appear, based on the pinnae and the rhizomes described as *Microlepiopsis*, to be a member of *Anemia*, subgenus *Anemiorrhiza*. Due to the present status of the fossil as *Microlepiopsis*, this placement is presently speculative and would need to be borne out with future studies. If truly related to the extant *Microlepia*, it would indicate moist wet environments but would also infer a subtropical to tropical climate. This fossil presently remains equivocal in its placement.

The pinnae alone show that *Anemia* was present in the flora, but, based on the remains of only the spore and pinnae, an extant form cannot be identified with any certainty.

The rhizome *Midlandia nishidae* is identified as a Blechnaceae with similarity of form to *Woodwardia*. There are up to twelve living species, but it is unknown which species would be appropriate. *Wessiea oroszii*'s affinities are less well known. Both ferns do not allow an accurate identification of a nearest living relative.

In the Dennstaedtiales with the suspected placement of the rhizomes *Microlepiopsis* in the *Anemia*, only isolated stipes remain. These cannot be identified to genera or species at this time along with the Gleicheniales, Cyatheales, and Ophioglossales.

In the past, the use of the conifer elements have been omitted in reconstructions of Cretaceous floras with emphasis placed on the already highly evolving angiosperms. The ancient conifer genera that have survived to the present still inhabit environmental settings that contain parts of the evolving angiosperm suite. These conifers are not only easier to identify and place into recent genera but are also easier to interpret environmentally.

The conifers that are easily identified or related to recent genera in the Horseshoe Canyon Formation include *Ginkgo*, *Taiwania*, *Metasequoia* (*Parataxodium*), *Taxodium*, *Cunninghamia*, *Torreya*, and *Athrotaxis*. Unfortunately the fossil *Picea* cannot be identified with any living species at this time, and the Caytoniales are extinct.

The seed cones previously interpreted as Cycadales (Serbet 1997) are interpreted here as Bennettitales. In the Horseshoe Canyon Formation, it is doubtful that seed cones representing the Bennettitales lack preserved foliage throughout the formation and vice-versa for the *Nilsonia* lacking reproductive structures.

The seed cones have been interpreted to be biovulate sporophylls and associated with the *Nilsonia-like* foliage (Serbet 1997), but unfortunately the seed cones are interpreted here as definitely cupulate. The foliage, without cuticular investigation, is best described as *Nilsonia*-like. If found to be a *Nilsonia* species, the placement of the Nilsoniales with the Bennettitales would be unique and would allow for the placement of the Nilsoniales and Cycadales to be questioned in other floras such as seed cones described by Kvaček (1997). The debate over the standing of the Horseshoe Canyon Formation seed cones and foliage as well as associates noted by others (Kvaček 1997) will no doubt be ongoing for many years.

Ginkgo, on the other hand, is easily identified and relatable to the extant *Ginkgo biloba*. *G. biloba* presently inhabits areas in close proximity to those of *Metasequoia*, *Cunninghamia*, and *Torreya*.

Taiwania is found from northern Yunnan to Szechwan and Hubei to the island of Taiwan. *Taiwania cryptomerioides* in mainland China is associated in areas with *Metasequoia glyptostroboides* present. It is predominantly below evergreen coniferous (*Abies*) belts and above deciduous forests. The mean annual temperature for Yunnan is 10°C with the mean coldest months ranging from 2–4°C (Wang 1961).

On the island of Taiwan, *Taiwania* is associated with *Cunninghamia konishii* and the hardwood *Trochodendron aralioides* in mixed, predominantly evergreen, forests. On the island of Taiwan, the mean temperature is higher, 11–15°C. Of the warmest months temperatures are 17.3 ± 2.4°C and on the coldest 7.6 ± 2.2°C (Wang 1961).

Different climatic factors are seen; warmer and wetter in Taiwan and drier and cooler in China. Taiwan forests are considered evergreen temperate, while forests on the Chinese mainland are considered to have a recognizable dry season. A maritime versus continental climate can be seen.

The direct living ancestor of *Parataxodium*, *Metasequoia glyptostroboides* grows in natural groves at latitude 30°10" N latitude and 108°35" W longitude. The forests in the area are considered as mixed mesophytic. The documented temperature range of this flora, in China, shows no mean month temperature below 0°C, although extreme minimum temperatures as low as −14°C are recorded (Wang 1961).

The *Metasequoia* flora itself has been well documented and exhibits floral composition rich in endemism and monotypic plants (Wang 1961; S. Y. Hu 1980). Many of the angiosperms present represent archaic families and genera, which are also found in the Horseshoe Canyon Formation such as *Rhus, Cornus, Cercidiphyllum, Sassafras, Fagus, Nyssa, Davidia, Betula, Toricellia, Carpinus, Ilex*, etc.

The conifers of the *Metasequoia* flora include *Cunninghamia, Taiwania, Torreya, Cupressus, Juniperus, Taxus*, and *Pinus*. A full listing of trees and shrubs from the *Metasequoia* area can be obtained from Wang (1961) or S. Y. Hu (1980).

The environment has been interpreted as primarily moist, mesophytic, and warm temperate in nature, and as a region of mild winters and hot summers with precipitation distributed throughout the year, but highest in the spring and summer. The *Metasequoia* area and surrounding forests are considered deciduous broadleaved forest (Wang 1961).

Taxodium distichum exists in both swamp and river locales. It grows in association with ferns such as *Osmunda, Onoclea*, and *Woodwardia* and in the vicinity of the rare conifer *Torreya taxifolia*. Some of the angiosperm flora represented in the fossil record, such as *Platanus, Nyssa* and *Sassafras*, also occur in these areas.

Specific fossil floras that contain *Taxodium* do not contain *Glyptostrobus*. In Alberta, *Glyptostrobus* is found in the Cretaceous, Dunvegan Formation. It does not make a re-appearance in the province until the Paleocene. Both extant genera live in riverine and pond environments, but extant *Glyptostrobus* is considered to grow in subtropical environments. Its last known range is in the southern areas of Fujian Province, China.

Cunninghamia lanceolata occurs in mountain valleys sporadically throughout south central China. Another species, *C. konishii*, is found on the island of Taiwan.

Torreya contains six species and grows both in Asia and North America. In China, *Torreya* is found in mixed woodland areas in the central and southeastern provinces that include areas where *Metasequoia, Taiwania*, and *Cunninghamia* occur.

Torreya in North America grows on riverbanks and valleys where similar archaic angiosperm groups occur along with the gymnosperm *Taxodium* in the southeast and *Sequoiadendron* on the west coast.

Both *Parataxodium* and *Cunninghamia* and all other wood types identified by previous authors from the Horseshoe Canyon Formation show growth rings. Living warm temperate conifers such as *Metasequoia, Taxodium, Cunninghamia,* and *Torreya* show distinctive rings similar to the fossils.

In the dicotyledonous angiosperms, the use of vegetation analysis based on leaf margins is common. This is based on the correlation between leaf margins and temperatures. Entire margin leaves decline in response to a decrease in temperature while non-entire margin leaves increase conversely.

Due to most of the angiosperm flora not being represented by actual leaf material, an interpretation using just them would be very biased. Many leaves that are present show non-entire leaves, but this may represent a bias due to the low number of leaf sites present in the formation and lack of substantial collections from these sites.

Some aquatic angiosperms that are considered subtropical such as *Pistia* and *Spirodela* are not found in the formation. Both *Pistia* and *Spirodela* are found in the Dinosaur Park Formation (74 mybp). They also occur in the early Paleocene of Saskatchewan. Both plants do occur in the stratigraphically equivalent St. Mary River Formation of southern Alberta, which lies 230 km to the south of Drumheller.

In contrast *Trapa*, in the form of *Trapago* or *Quereuxia* are found in the Dinosaur Provincial Park, Horseshoe Canyon, St. Mary River, and Ravenscrag floras. This may reflect a temperature tolerance greater than that exhibited by either *Pistia* or *Spirodela*. Extant *Trapa* is considered warm temperate to subtropical in distribution.

The fossil *Albertarum pueri* is related most closely to extant *Symplocarpus* (subfamily Orontioideae), the Skunk Cabbage. *Symplocarpus* is found in both eastern North America and Asia. It prefers edges of swamps and wet ground in warm temperate climates.

Even with reproductive structures such as seeds and fruits, it can be seen that the angiosperms were still rapidly evolving, based on the lack of fossil forms that can be ascribed to living genera and species. Of the identified angiosperm genera relatable to modern genera found in the formation, many also occur with the Taxodiaceae genera in modern floras. These are again somewhat restricted to eastern China and the eastern, western and southern coasts of the United States, although some have ranges far beyond this (i.e., Platanus, Carpinus, etc.). This may reflect the ability of these genera to adapt more quickly to different environments while still colonizing ancestral environments. The angiosperms as a whole would seem to indicate a warm temperate to subtropical environment. Wholly tropical forms are lacking.

Fig. 802.   Fossil Trichoptera larval casing.

ENVIRONMENTAL RECONSTRUCTION OF THE HORSESHOE CANYON FORMATION BASED ON THE FAUNA

Climatic conditions based on the fauna do give an indication of environment similar to the plants, although a large amount of the taxa are extinct. All dinosaurs, save the living relatives, birds, are extinct, as are the large marine reptiles. Mammals found in the formation are basal lineages and habitat preferences are unknown. Frogs and salamanders are known to occur in the fossil record of the formation, but they are presently unstudied.

Trichoptera (Caddis flies), although distributed worldwide, are species specific to environments and foraging habits. Each species also has specific pupal case construction. These are rare fossils with only two specimens found to date. Both fossil specimens are similar in construction and consist of small cemented mud pellets of uniform size (Fig. 802). The fossil casings would appear to belong to the Limnephilidae, the Northern Casemaker Caddisfly Family, in the subfamily Neophylacinae that contains four genera: *Farula*, *Neophylax*, *Neothremma*, and *Oligophlebodes*.

The pupal case is most similar to those constructed by the genus *Farula* that contains seven living species, which range from warm temperate to subarctic.

Many gastropod species are present in the formation such as:

Order Archaeogastropoda
    Family Naticacea (Moon Snails)
        Polinices (Euspira) obliquata
        Polinices euspira occidentalis
Order Caenogastropoda
    Family Viviparidae (Mystery Snails)
        Campeloma sp.
        Viviparus prudentius
        Viviparus tasgina
        Viviparus westoni
    Family Valvatidae (Valve Snails)
        Valvata filosa
    Family Pilidae (Apple Snails)
        Reesidella sp. cf. *R. protea*
    Family Pleuroceridae (River Horn Snails)
        Campeloma edmontonensis
        Goniobasis sp.
        Goniobasis webbi
        Lioplacodes limnaeformis
        Lioplacodes sanctamariensis
        Lioplacodes cf. *L. tenuiscarinata*
        Lioplacodes whiteavesi
Order Archaeopulmonata
    Family Ellobiidae
        Pleurolimnaea mclearni
Order Branchiopulmonata
    Family Physidae (Pond Snails)
        Physa sp.

Fossil gastropods are found in a variety of preservational states from silica- or carbonate-replaced to mud in-fills (Figs. 803, 804). Unfortunately the related extant taxa show a wide range in

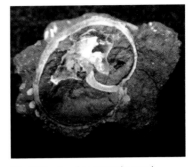

Fig. 803. Fossil gastropod natural fractured cross-section.

Fig. 804. Fossil gastropod partially prepared.

habitat. The fresh water *Viviparus* and *Goniobasis* are both gastropod types of warm temperate as well as tropical climates.

A large number of bivalve species representing both fresh and salt water forms are found. These are:

Order Nuculoida
    Family Nuculidae (Nut Clams)
        Nucula subplana
Order Arcoida
    Family Cucullaeidae
        Cucullaea shumardi
Order Unionida
    Family Unionidae (Freshwater Mussels)
        Fusconaia? danae
        Lampsilis? consuetus
        Rhabdotophorus senectus
        'Unio' albertensis
        'Unio' minimus
        'Unio' priscus
        'Unio' sandersoni
        'Unio' stantoni
        Plesielliptio sp.
Order Veneroida
    Family Pisidiidae (Fingernail Clams)
        Sphaerium heskethense
Order Pterioida
    Family Anomiidae (Jingle Shells)
        ?Anomia micronema
        Anomia perstrigosa
    Family Ostreidae (Oysters)
        Crassostrea subtrigonalis
Order Mytiloida
    Family Mytilidae (Sea Mussels)
        Brachidontes dichotoma
        ?Crenella elegantula
        Modiolus galpinanus

Modiolus meeki
Mytilus albertensis
Order Veneroida
    Family Corbiculidae
        Corbicula cytheriformis
        Corbicula occidentalis
        Corbicula occidentalis ventricosa
    Family Veneridae (Venus Clams)
        Callista deweyi
        Callista nebraskensis
Order Myoida
    Family Corbulidae (Box Clams)
        Corbula subtrigonalis
        ?Corbula perangulata
    Family Hiatellidae
        Mya simulatrix

Salt water forms may prove useful to indicate near-shore environments but studies from this area are unknown. Freshwater fossil *Unio* shells are locally abundant and are commonly found as ironstone in-fills called *steinkerns* (Fig. 805). Living unionioids typically inhabit well-oxygenated, shallow, calcium-rich, moving fresh waters that are seasonally warm and have a pH greater than 7. The Red Deer River has unionioids living in it today. The clam fauna exhibits a wide range of habitat preference.

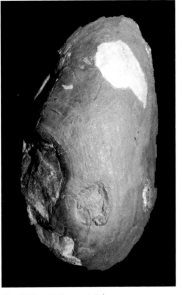

Fig. 805. *Unio* sp. steinkern.

The fish fauna contains many genera with extant counterparts. A listing of the fish in the formation includes:

Class Chondrichthyes (Cartilaginous Fishes)
    Order Rhinobatiformes (guitar fish)
        Family indet.
            Myledaphus bipartitus Cope, 1876
    Order Lamniformes (Sharks and Rays)
        Lamna sp.

Class Osteichthyes (Higher Bony Fishes)
    Order Acipenseiformes
        Family Acipenseridae (Sturgeons)
            Acipenser sp.
            Diphyodus longirostris
        Family Polyodontidae (Paddlefish)
            Palaeopsephurus sp.
    Order Lepisosteiformes
        Family Lepisosteidae (Gars)
            Lepisosteus sp.
    Order Amiiformes
        Family Amiidae (Bowfin)
            Cyclurus fragosus
    Order Aspidorhynchiformes
        Family Aspidorhynchidae
            Belonostomus sp.

Fig. 806. *Myledaphus* sp. tooth.

The ray fish, *Myledaphus*, is presently placed in the order Rhinobatiformes, which are closely related to the ray fish. It is not placed in any of the three extant families of Rhinobatids as its relationships are inconclusive. The fossil teeth are common in freshwater and brackish water deposits in the formation (Fig. 806).

Acipenser sp. and *Diphyodus longirostris* belong to the Acipenseridae (Sturgeons), which occur in near-shore oceanic as well as lake and river environments almost worldwide. There are twenty-seven species of sturgeon; nine are endemic to North America. It is unknown what would be the nearest living relative to the fossil form, although two species (*A. sinensis*, *A. dabryanus*) occur in the Yangtze River drainage and one (*A. transmontanous*) in the Mississippi River drainage.

Palaeopsephurus sp. is in the family Polyodontidae (paddlefish), which only contains two extant species *Polydon spathula* and *Psephurus gladius*. *Polydon* lives in the Mississippi River drainage of

North America. Its most northerly range is recorded from southern Ontario, Canada. *Psephurus* is restricted to the upper Yangtze River drainage of China. Both prefer quiet, slow-moving water habitats. Both fish exist in areas where taxa of the ancient flora exist.

The family Lepisosteidae (Gars) is represented by the genus *Lepisosteus* with four species and *Atractosteus* with three species. All are restricted to North America, Central America, and Cuba. It is unknown which genus is represented by the fossil species in the formation.

Amia calva is the nearest living representative for the fossil fish, *Cyclurus fragosus*. *Amia calva* is restricted to North America from Texas through the Mississippi River drainage and eastern seaboard to Ontario and Quebec, Canada.

The *Belonostomus* is an extinct fish genus from the formation.

The turtle fauna is much reduced when compared to other coeval formations further south and consists of the following:

Trionychidae
    Aspideretoides sp. cf. *A. foveatus*
Nanhsiungchelyidae
    Basilemys sp.
Adocidae
    Adocus sp.
Macrobaenidae
    gen. et sp. indet.
Chelydridae
    gen. et sp. indet.
?Eucryptodira gen. indet.

The turtle fauna of the formation and their paleo-ecological interpretations have been studied (Brinkman 2003; Brinkman and Eberth 2006) with the result that turtle diversity correlates strongly with paleo-temperatures (Brinkman and Eberth 2006). The rarity of turtles in the formation was assumed to have a paleo-ecological rather than a taphonomic bias. A decrease in turtle diversity during the Campanian/Maastrichtian documented by Brinkman and Eberth (2006) also correlates with a cooling trend noted by Upchurch and Wolfe (1993) based on contemporaneous floras within North America.

Champsosaurs are also found in the formation. The Champsosauridae are a family of aquatic reptiles that became extinct in the middle Eocene. *Champsosaurus*, with a long snout lined with teeth, resembled modern Gharials, although they are not related. Their similar physical appearance represents convergence of form based on feeding habits. Both extinct Champsosaurs and modern Gharials feed on fish.

Isolated "Croc" teeth commonly found in the formation, in the past, have been attributed to the Crocodylidae but represent those of the Alligatoridae based on recovered skeletons (*Stangerochampsa*, Wu et al. 1996). The nearest occurrence of the Crocodylidae is in the upper Scollard Formation (earliest Paleocene) and Dinosaur Park Formation (Cretaceous), both of which represent differing ages and different paleo-environments.

Present-day Alligatoridae are found in both China and North America. Chinese alligators are restricted to the lower Yangzi River, Anhui Province. They prefer slow-moving to still waters in ponds and rivers and feed extensively on a diet of hard-shelled molluscs.

American alligators live in the southeastern United States from Texas across to North Carolina and feed on all aquatic and terrestrial prey.

American alligators can hibernate during winter months. Chinese alligators live in warm temperate climates, while those in America are warm temperate to subtropical in distribution. American alligators have been known to swim in freezing waters under ice in the northern limit of their range.

The use of fauna alone gives a wide range in climate forms, although some of the taxa such as the Alligatoridae, turtles and certain fish genera are somewhat more climate restricted. Both Asian and North American Alligatoridae and certain fish species exist in areas that contain plant genera found in the fossil flora.

Some fossil plant sites have also produced specific faunas associated with the flora. These are:

Kent's Knoll (An Allochthonous Site)
    Albertosaurus sp. shed teeth
    Hadrosaur sp. shed teeth
    Small Theropod sp. Indet. shed teeth
    Champsosaurus sp. teeth and vertebrae
    Alligatoridae sp. shed teeth
    Unio sp. steinkerns
    Myledaphus sp. shed teeth
    Diptera fam. indet. Puparia

Day Digs (An Autochthonous Site)
    Viviparus sp. shells and opercula
    Goniobasis sp. shells and opercula
    Unio sp. steinkerns
    Trichoptera gen. Larval casings

Cemetery site (An Autochthonous Site)
    Unio sp.
    Hadrosaur sp.

FOOD FOR THOUGHT: DINOSAUR/PLANT INTERACTIONS

There are many questions that are asked about dinosaur and plant interactions of the Upper Cretaceous. Below are some differing interpretations to answer some of the more common questions such as: what did the herbivorous dinosaurs in the formation eat?

First, before one can answer what the dinosaurs ate, one must know where the dinosaurs actually lived. Although many dinosaur skeletons are found in deltaic and riverine deposits, it is highly unlikely that they actually lived there. The adult herbivorous dinosaurs found in the Horseshoe Canyon Formation such as hadrosaurs or ceratopsians possibly weighted 2–4 tonnes.

Hadrosaurs were herding animals that stayed in groupings of similar-sized animals. Conversely, ceratopsians were herding animals as well with many different-sized individuals making up the herd. These two herding strategies are documented in fossil bonebeds found in North America. These animals did not necessarily live by the river in which their bones were deposited. They merely may have had to cross it at an inopportune time, perhaps during migration. A bloated carcass of a 2–4 tonne animal can be easily transported by a large river in flood stage and individual bones can be transported far from their source.

Bonebeds are examples of bones transported over distances as the animals died en masse in a catastrophic event. The carcasses were either scavenged or left to decay in the river floodplain. These body parts were later moved and sorted during another catastrophic event or events, which led to their eventual burial downstream. Many bonebeds attest to the moving and size-sorting of bone elements, which can occur over long distances. Many recent analogies of large herding animals caught in catastrophic flood events have been documented.

Due to the dinosaur's weight and herding strategies, it is highly unlikely these large animals would venture into heavily forested or swamp-like environments. Ceratopsians also had the added disadvantage of large bone frills attached to their skulls, which would likely be a hindrance in heavily forested areas. These were animals that most likely inhabited more open, seasonally stable environments.

WHERE ARE THESE SEASONABLY STABLE ENVIRONMENTS PRESERVED?

The majority of fossil plant sites are the result of catastrophic events. If they represented seasonal changes in stable environments, there would be millions of fossil plant sites but there are not. The exceptions to this rule are coals, which represent quiet-water accumulations. Even many of these are truncated by river sediments and may contain catastrophic events in the way of fire burn or unseen disconformities due to expulsion of plant materials during flooding. Coals do not necessarily represent yearly seasonal accumulations as they are more a positive net gain accumulation over time.

All individual plant sites represent some catastrophic event in the story of delta subsidence and infill such as storm surges in the delta, sporadic spring storms, cyclic fire burn, river channel flood stage changes, swamp succession, and Fall burial of deciduous leaves in quiet backwaters.

Unfortunately, seasonably stable environments are rarely preserved. Hadrosaur nesting sites of the southern Alberta Oldman Formation are one example of preserved seasonably stable environments by sporadic catastrophic events. The fossilization of eggs in these areas is considered exceptional and rare due to sporadic catastrophic events that occurred in these environments. It is only through events such as a levee breach, massive mud flow, or volcanic disturbance that such environments are preserved. Conversely, living too close to deltas or close to large river systems increases the seasonal remodelling of potential living areas, making it less habitable to large herding herbivores. Why nest there if there is a greater potential for seasonally catastrophic events? The species would soon become extinct.

WHAT PLANTS ARE FOUND IN THESE SEASONABLY STABLE ENVIRONMENTS?

Plant remains in these areas are scare and rarely preserved similar to the dinosaurs. Based on preliminary investigations of plant remains in nesting sites in the Oldman Formation of Alberta, the flora appears unlike that of the preserved delta areas (Horseshoe Canyon Formation). Unfortunately, many of the plant remains have not been sufficiently investigated to quantify the differences. Preliminary investigation of these environments shows the potential to introduce many new plant taxa as many new forms of angiosperms, gymnosperms, and ferns appear present.

The Oldman Formation is also older geologically than the Horseshoe Canyon Formation. In Alberta, coeval seasonally stable environments to the Horseshoe Canyon Formation are not known. Without a coeval environment to the Horseshoe Canyon Formation, it is unknown what plant types would have occurred in the more stable environments of the time.

CAN WE TELL WHAT AN HERBIVOROUS DINOSAUR ATE BY ITS TEETH OR STOMACH CONTENTS?

Herbivorous dinosaurs had exceptionally specialized teeth for grinding their food source. These tooth forms had evolved over millions of years and most likely in conjunction with specific vegetation. We would need to look back to the Triassic, Jurassic, or Lower Cretaceous floras and plants available to start searching for common occurrences of plants then and later on in the Upper Cretaceous. Many of the Filicophytes, Cycadophytes, Ginkgophytes, and Coniferophytes follow the evolution of the dinosaur families.

Presently the study of teeth in various herbivorous dinosaur groups has looked at compacted plant material between teeth as well as microscopic wear patterns. These studies have not yet been published. Scrapings of teeth would need to be undertaken on "mummified" specimens as those exposed to the elements during burial in a fluvial environment may have plant material compacted against the tooth rows or batteries giving false results. As for wear patterns, unfortunately many Cretaceous fossil plants are extinct and their ability to produce specific wear patterns are conjectural or unknown.

Only a few rare mummified skeletons of hadrosaurs have been examined for stomach contents. Dinosaurs are not truly "mummified." The term mummified, when used to describe dinosaur finds, refers to the preservation/fossilization of soft tissues such as skin, nails, and in very rare cases internal tissues such as the digestive tract or muscle in a semi-three-dimensional desiccated state. Many "mummified" dinosaur remains consist of the tougher skin preserved as an imprint infilled with sediments similar to a glove filled with sand. There are presently only a few dozen rare dinosaur fossils worldwide that preserve softer internal tissues such as intestine and muscle. Most, if not all, of these represent small-sized carnivores. Individual bones and partial carcasses of herbivores found in fluvial environments do not preserve internal tissues. Present claims of such finds are considered a misinterpretation of the fossil and its preservational state.

One should be wary of the interpretation of mummified fossilized remains. First, any stomach contents exposed outside the body cavity may not be stomach contents but merely sorted plant hash that accumulated in the open body cavity during burial. It is doubtful that a large herbivore would desiccate prior to fossilization without its stomach contents rupturing into the surrounding environment as well as the environment entering and contaminating the stomach cavity.

Questions that would need to be answered are things such as: is the plant material restricted to the abdomen of the animal or also in the rib cage (lungs) or disseminated outside the body wall? As an example, one hadrosaur at the Royal Tyrrell Museum contained a fish skeleton in the rib cavity along with large amounts of plant material. The hadrosaur did not eat the fish. The fish swam into the carcass prior to burial and was somehow trapped along with the plant material. This is not an isolated occurrence. Burial in a dertrital environment may give a false indication of stomach contents.

If the mummified skeleton shows no such rupturing to the outside environment, stomach contents may be found. The question now asked should be: do the stomach contents in such an emaciated animal then represent a healthy individual and its normal food source prior to death or are they the remnants of what a starving individual would have tried to eat prior to death? This could really only be answered with a larger sample size rather than just one or a few individual animals. These studies are ongoing.

SO WHAT DID THE HERBIVOROUS DINOSAURS EAT?

It is suspected that the dinosaurs did not feed upon the evolving angiosperms and that they were already specialized feeders on the previously mentioned vegetation sources with which they evolved. Some have suggested that the ceratopsians and hadrosaurs fed on angiosperm remains, based on skeletons found with associated wood flora (Lehman and Wheeler 2001). This merely represents depositional environment, not feeding strategies. Hadrosaurs found in the Horseshoe Canyon Formation most likely feed on specific, large shrub or tree Ginkgophytes and Coniferophytes. The ceratopsians were most likely low browsers on specific Filicophytes or Cycadophytes similar to other low browsers such as ankylosaurs.

WHAT KILLED THE DINOSAURS?

There is the common misconception in the general public that the dinosaurs became extinct during an extraterrestrial impact at the end of the Cretaceous. This impact was real as the Cretaceous/ Tertiary boundary is found around the world. It marks the line where there are dinosaurs below, but none above. But is this really what caused the end of the dinosaurs, or is it merely a marker of a catastrophic event of worldwide proportions, which did not affect an already extinct life form?

There are presently no known dinosaur skeletons or bones found at the Cretaceous/Tertiary boundary. Why is this? If this were their end, would there not be some sort of evidence at the boundary? There are a few theories such as caustic minerals in the impact layer which dissolved calcium in the bones or coal-seam contacts either above or below the layer again leaching out calcium. Unfortunately, many dinosaur bones can be found in contact with coals or coaly shales in this and other formations. Is there not enough exposure of the Cretaceous/Tertiary boundary around the world in which to look for skeletal remains?

In the province of Alberta, the nearest articulated skeleton to the boundary is a *Triceratops* sp., three metres below. This sediment can represent tens to hundreds of thousands of years. Although there are some reports of individual bones found in sediments above the boundary, it is not uncommon for rivers to carve many metres down into underlying sediments and pull up individual bones for re-deposition. This occurs even in today's environments.

The time of the *Tyrannosaurus Rex*/*Triceratops* faunas are considered to be the pinnacle of evolution in many of the dinosaur lines. This specialization may have come at a high price. If the dinosaurs did evolve to consume the more primitive plant forms, their digestive tracts may have been ill-equipped to process angiosperms.

The first angiosperms were thought to have been herbaceous and shrub-like in form. If these forms were anything like some forms present today, they most likely reached reproductive maturity fairly quickly (5–10 years). It is well known that many conifers today take twenty to forty years to reach reproductive maturity. When dinosaurs ate their food source in a particular area, the more advantageous, quicker-growing angiosperms may have taken their place. This early competition may have set the scene for dinosaur demise.

By the Upper Cretaceous, tree forms of angiosperms are known to have occurred (Lehman and Wheeler 2001; Wheeler and Lehman 2000; Wheeler et al. 1994). This evolution may have been a response to ecological niches left open by dwindling Coniferophyte and Ginkgophyte forests. With palatable food sources scarce, dinosaurs may have been forced into longer migrations to find food. When a specific food source could no longer be found, the dinosaur linked to this source would die out. The predators that relied on these particular dinosaurs would also be under greater stress and would die off if a suitable large herbivore could not be found as a replacement.

The plant food source need not become extinct; it just needs to be isolated by distance or geography from its herbivore. If a mountain range or river system should isolate a plant from its forager, unless the forager can find a suitable palatable substitute or find another route to the food source, it will become extinct.

The landscape of the Upper Cretaceous was changing rapidly in North America. The shallow inland sea was disappearing, coastal mountain building was underway, and the land was cooling and drying. Other Cretaceous environments worldwide were also undergoing similar changes. All these changes, together with the spread of angiosperms, may have condemned the dinosaurs to a slow but steady extinction over millions of years, only to be postmarked later by a worldwide catastrophic event.

CONCLUSION

The variety of plant types found in the Horseshoe Canyon Formation has increased substantially since W. A. Bell's 1949 report; unfortunately, little still has been published scientifically. Even so, many preliminary conclusions can be drawn from the flora.

The Cretaceous flora of the Horseshoe Canyon does not substantiate a tropical to subtropical claim at the present. The majority of evidence places the climate as being humid warm temperate to warm temperate with distinct seasonality. Severe frost may have occurred such as experienced by areas presently occupied by the living floras of central China and southeastern North America.

Although many of the extant families are present in the formation's flora, few represent modern genera and none represent modern species, based on this author's investigation.

The Taxodiaceae were a dominant component of the flora, and the angiosperms that were present appear mainly as deciduous forms. Although this conclusion may presently be biased towards the collections available at this time, based on the northerly position of the area during the Cretaceous and the lower "winter" seasonal light regime, the flora seems to be consistent with a predominantly deciduous, humid, warm temperate climate model.

As for specific components of the flora, the fern flora is presently being scientifically investigated with hopefully many new insights and discoveries to be made. A major question for inquiry is: where are the tree fern remains? Are they present but with mega-plant preservational bias or is there some other reason for their absence? Do the spore reflect true tree fern spore?

What do some of the other fern families represent in the form of genera or species?

The conifers, although fairly well-known, now require interpretation and further evolutionary studies. Do the Bennettitales seed cones give us a last glimpse at the end of an evolutionary line of specialization? Is the Alberta Maastrichtian the last occurrence of this once widespread order or did it survive into the Paleocene?

Does true *Metasequoia* in its present form exist anywhere in the Cretaceous worldwide or does it evolve from *Parataxodium* only in the early Paleocene and why? Is the *Elatides* species present a precursor to *Taiwania* or is it merely an extinct relative? What do the *Athrotaxis* species seed cones look like? What pine genera can we hope to find in the future?

The angiosperms were still a rapidly evolving group. The canopy and shrub flora appears adequately represented, but this is based mainly on palynomorphs. Old and new seed and leaf sites need to be described. The lower ground covers in the form of herbaceous dicots and monocots, although now present, is poorly represented or investigated. What do these plants look like? Are there any evergreen genera?

Rare plant forms such as monocot remains are becoming more prevalent in the formation. This is attributed to our understanding of which parts may fossilize and how to identify them in the field.

Many new rhizome forms have been found, but morphological and evolutionary studies are lacking for accurate comparison with extant taxa.

This area promises to be one of the most rewarding for future studies.

The unknown palynomorphs in all orders hint at many more discoveries to be made and many more questions to be asked.

The Upper Cretaceous Horseshoe Canyon Formation contains an untapped wealth of information on floral evolution. It is hoped that this book garners interest in either fossils found or yet to be found by the general public and the scientific community.

There is still much to be discovered and written on the flora of the Horseshoe Canyon Formation of Drumheller, Alberta.

GLOSSARY

Abaxial *adj.* Located away from or on the opposite side of the *axis*; *dorsal*. Opposite of *adaxial*.

Aberrant *adj.* Deviating from what is normal; untrue to type.

Abscission *n.* The shedding of stems, leaves, flowers or fruits following the formation of the *abscission* zone.

Abundant *adj.* Occurring in or marked by abundance; plentiful.

Achene *n.* A small dry indehiscent fruit with a thin wall.

Achlorophyllous *adj.* Having no chlorophyll.

Acidic *adj.* Being or containing an acid; of a solution having an excess of hydrogen atoms (having a pH of less than 7).

Adaxial *adj.* Located on or nearest to the *axis*; *ventral*. Opposite of *abaxial*.

Affinity *n.* A relationship or resemblance in structure between species that suggest a common origin.

Allochthonous *adj.* Originating from outside a system, such as the leaves of terrestrial plants that fall into a stream.

Alluvial *adj.* Of, relating to, or found in sediments deposited by flowing water (alluvium).

Amorphous *adj.* Lacking definite form; shapeless.

Amphipholic *adj.* Having *phloem* on both sides of the *xylem*.

Amphisporangiate *adj.* Containing both mega- and microsporangia in the same *sorus*, used to describe certain sori in *heterosporous* ferns.

Amphivasal vascular bundle *n.* A concentric vascular bundle in which the *xylem* surrounds the *phloem*.

Anemophilous *adj.* Pollinated by wind-dispersed pollen.

Angiosperm *n.* A plant whose ovules are enclosed in an ovary; a flowering plant.

Annual *adj.* Living or growing for only one year or season. *n.* A plant that completes its entire life cycle in a single growing season.

Annulated *adj.* Having or consisting of rings or ring like segments.

Annulus *n.*; **annuli** or **annulues** *n. pl.* A ring or group of thick-walled cells around the *sporangia* of many ferns that functions in *spore* release.

Antheridiophore *n.* Stalk bearing antheridia as in certain mosses and liverworts.

Anticlinal wall(s) *n.* The cell wall perpendicular to the surface of a plant organ.

Aperture(s) *n.* An opening, such as a hole, gap, or slit.

Apical *adj.* Of, relating to, located at, or constituting an apex.

Arborescent *adj.* Having the size, form, or characteristics of a tree; treelike.

Archaic *adj.* Of, relating to, or characteristic of a much earlier, often more primitive, period. No longer current or applicable.

Archegoniophore *n.* Stalk bearing archegonia as in certain mosses and liverworts.

Arecoid *adj.* Resembling, having the appearance of or related to the Arecaceae.

Arenchyma *n.* A spongy tissue with large air spaces found between the cells of stems and leaves of certain plants.

Arillate *adj.*; **aril** *n.* Of some seeds; having a fleshy and usually brightly coloured cover arising from the hilum or funiculus.

Aromatic *adj.* Having an aroma; fragrant or sweet-smelling.

Autochthonous *adj.* Originating or formed in the place where found.

Axil *n.* The upper angle between a lateral organ, such as a leaf stalk, and the stem that bears it.

Axis *n.* The main stem or central part about which organs or plant parts such as branches are arranged.

Basal *adj.* **a.** Of, relating to, located at, or forming a base. **b.** Located at or near the base of a plant stem, or at the base of any other plant part.

Bennettitales *n.* An *extinct* order of fossil gymnospermous plants.

Bifacial *adj.* Having two fronts, faces, or facades.

Bifurcate *v.* To divide into two parts or branches.

Bifurcation *n.* The place where something divides into two parts or branches.

Bijugate adj. Relating to a pinnate leaf with two pairs of leaflets.

Biotite *n.* A dark-brown or dark-green to black mica found in igneous and metamorphic rocks.

Bipinnate *adj.* Of a leaf shape; having doubly *pinnate* leaflets (as ferns).

Biseriate *adj.* Arranged in two rows or in two cycles.

Blade *n.* The expanded part of a leaf or petal.

Bract *n.* A leaflike or scalelike plant part, usually small, sometimes showy or brightly coloured, located just below a flower, a flower stalk or an *inflorescence*.

Bud scales *n.* A scaly leaf that is part of a protective sheath around a plant bud and is sometimes hairy or resinous.

Bulb *n.* A short modified underground plant part that consists of fleshy modified leaves that contain stored food for the shoot within, attached to a much-reduced stem and roots, i.e., onion *bulb,* tulip *bulb.*

Calcification *n.* The hardening of a structure, tissue, etc., by the *deposition* of salts of lime, as in the formation of teeth, and many forms of *petrification.*

Campanian *n.* The second last stage of the Cretaceous occurring from 83.5 ± 0.7 Ma to 70.6 ± 0.6 Mybp.

Carbonization *n.* Conversion into carbon, charcoal, or coke.

Carcass *n.* The dead body of an animal, especially one slaughtered for food.

Carpel *n.* One of the structural units of a pistil, representing a modified, *ovule*-bearing leaf.

Chaotic *adj.* A condition of great disorder or confusion.

Circumferentially *adj.* Lying around or just outside the edges or outskirts.

Clade *n.* A group of organisms, such as a species, whose members share homologous features derived from a common ancestor.

Cladistics *n.* A method of classifying living organisms, often using computer techniques, based on the relationships between phylogenetic branching patterns from a common ancestor.

Climatology *n.* The meteorological study of climates and their phenomena.

Clone *n.* An organism descended asexually from a single ancestor, such as a plant produced by layering or budding of a *rhizome.*

Coeval *adj.* Originating or existing during the same period; lasting through the same era.

Collateral vascular bundle *n.* A bundle having *phloem* only on one side of the *xylem,* usually the *abaxial* side.

Colonize *v.* To form or establish a colony.

Colpus *n.* **Colpi** *n. pl.* An elongated, aperture with a length/breadth ratio greater than 2.

Commensalistic *adj.* To be in a symbiotic relationship between two different organisms when one receives benefits from the other without damaging it.

Compression *n.* The state of being compressed.

Configuration *n* Arrangement of parts or elements.

Consolidate *v.* To make strong or secure; strengthen.

Corm *n.* A short thick solid food-storing underground stem, *rhizome.*

Corrugated *v.* Shaped into folds or parallel and alternating ridges and grooves.

Cortex *n.* The region of tissue in a root or stem lying between the *epidermis* and the vascular tissue.

Cosmopolitan *adj.* Growing or occurring in many parts of the world; widely distributed.

Cotyledon *n.* A leaf of the *embryo* of a seed plant, which upon germination either remains in the seed or emerges, enlarges and turn green. Also called a *seed leaf.*

Crassula *n.;* **Crassulae** *n. pl.* Thickening of intercellular material and primary wall along the upper and lower margins of a pit-pair in the *tracheids* of *gymnosperms.* Also called *bars of Sanio.*

Crenulated *adj.* Having a margin with very small, low, rounded teeth.

Crosier *n.* The coiled young *frond* of any of various ferns. Also called a *fiddlehead*.

Cryptogam *n.* A member of a formerly recognized taxonomic group that included all seedless plants and plant-like organisms such as mosses, algae, ferns, and fungi.

Cupressoid *adj.* Resembling, having the appearance of or related to the Cupressaceae.

Cupule *n.* A small cup-shaped structure or organ, such as the cup at the base of an acorn.

Cuticle *n.* The continuous film of *cutin* on certain plant parts.

Cutin *n.* A mixture of fatty substances found in *epidermal* cell walls and in the *cuticle* of plant leaves and stems.

Cycadales *n.* An order of tropical primitive *gymnosperms* abundant in the Mesozoic, now reduced to a few scattered tropical forms.

Debris *n.* The scattered remains of something broken or destroyed; the fragmented remains of dead or damaged cells or tissue.

Deciduous *adj.* Shedding or loosing foliage at the end of the growing season.

Decurrent *adj.* Having the leaf base extending down the stem below the insertion.

Decussate *adj.* Arrangement of leaves in pairs which alternate with one another at right angles.

Degradation *n.* The act or process of degrading.

Departing trace *n.* The vascular strand or strands, which leave the vascular cylinder but are still within the stem, prior to entering the leaf or lateral organ.

Deposition *n.* The act of depositing, especially the laying down of matter by a natural process.

Dermal *adj.* Of or relating to the dermis layer below the *epidermis*.

Desiccate *v.* To dry out thoroughly.

Detritus *n.* Accumulated material; debris.

Deviate *v.* To turn aside from a course or way.

Devoid *adj.* Completely lacking.

Diagnostic *adj.* Serving to identify a particular characteristic.

Diarch *adj.* Primary *xylem* of the root; having two *protoxylem* strands, or two *protoxylem* poles.

Dichotomous *adj.* Divided or dividing into two parts or classifications.

Dichotomy *n.* Branching characterized by successive forking into two approximately equal divisions.

Diclinous *adj.* Having stamens and pistils in separate flowers.

Dicotyledonae *n.* A class comprising seed plants that produce an *embryo* with paired cotyledons and net-veined leaves.

Differentiated *adj.* Made different (especially in the course of development) or shown to be different.

Diminutive *adj.* Extremely small in size; tiny.

Dimorphic *adj.* Existing or occurring in two distinct forms; exhibiting dimorphism.

Dioecious *adj.* Having the male and female reproductive organs borne on separate individuals of the same species.

Diploid *adj.* Having a pair of each type of chromosome, so that the basic chromosome number is doubled.

Disperse *v.* To drive off or scatter in different directions.

Dispose *v.* To place or set in a particular order; arrange.

Disposition *n.* Arrangement, positioning or *distribution*.

Dissection *n.* Cutting so as to separate into pieces.

Dissolution *n.* Decomposition into fragments or parts; disintegration.

Distribution *n.* The geographic occurrence or range of an organism.

Divergence *n.* The act of moving away in a different direction from a common point.

Dormant *adj.* In a condition of biological rest or inactivity characterized by the cessation of growth or development and the suspension of many metabolic processes.

Dorsal *adj.* Of or on the outer surface, underside or back of an organ.

Dorsoventrally *adv.* In a dorsoventral direction; back to front.

Drupe *n.* A fleshy fruit, such as a peach, plum, or cherry, usually having a single hard stone that encloses a seed. Also called a *stone fruit*.

Ellipsoidal *adj.* Having a geometric surface, all of whose plane sections are either ellipses or circles.

Elliptical *adj.* Of, relating to, or having the shape of an ellipse.

Embryo *n.* Minute, rudimentary plant contained within a seed or an archegonium.

Endemic *adj.* Prevalent in or peculiar to a particular locality or region.

Endemism *n.* Nativeness by virtue or originating or occurring naturally (as in a particular place).

Endocarp *n.* The innermost layer or layers of the pericarp.

Endocast *n.* A copy or cast of the inside of an object.

Endodermis *n.* The layer of ground tissue forming a sheath around the vascular region and having the casparian strip in its *anticlinal* wall; may have secondary walls later. It is the innermost layer of the *cortex* in roots and stems of seed plants.

Enigmatic *adj.* Of or resembling an enigma; puzzling.

Ephemeral *adj.* Lasting for a markedly brief time; temporary.

Epidermal *adj.* Of or relating to a *cuticle.*

Epidermis *n.* The outer layer of cells in the plant body primary in origin. If it is multiseriate (multiple *epidermis*), only the outermost layer differentiates as the *epidermis.*

Epiphyte *n.* A plant that grows on another plant upon which it depends for mechanical support but not for nutrients.

Epithelial *adj.* Of or belonging to an *epithelium.*

Epithelium *n.* A compact layer of cells, often secretory in function, covering a free surface or lining of a cavity.

Equivocal *adj.* Open to two or more interpretations and often intended to mislead; ambiguous.

Erroneous *adj.* Containing or derived from error; mistaken.

Estuarine *adj.* Of, relating to, or found in an estuary.

Eudicot *n.* Short form of **Eudicotyledonae** *n.* A major group of non-magnolid dicots, containing *triaperaturate* or *triaperaturate* derived pollen.

Exine *n.* The outer layer of the wall of a spore or pollen grain.

Exocarp *n.* The outermost layer or layers of the pericarp.

Exposure *n.* Geological sediments exposed via erosional processes.

Extant *adj.* Still in existence, not destroyed, lost, or *extinct.*

Extinct *adj.* No longer existing or living.

Extinction *n.* The fact of being *extinct* or the process of becoming *extinct.*

Facial *adj.* Of or concerning the face.

Falcate *adj.* Curved and tapering to a point; sickle-shaped.

Fascicled *adj.* Growing in a bundle, tuft, or close cluster.

Fibro-vascular bundle *n.* A strand of the vascular system in stem and leaves of higher plants consisting essentially of *xylem* and *phloem.*

Filiform *adj.* Threadlike.

Floats *n.* An air-filled sac or structure that aids in the flotation of an aquatic organism.

Fluvial *adj.* Of, relating to, or inhabiting a river or stream.

Formation *n.* The primary unit of lithostratigraphy consisting of a succession of strata useful for mapping or description.

Fossilization *v.* The action or process of fossilizing; the conversion of vegetable or animal remains into fossils.

Friable *adj.* Readily crumbled; brittle.

Frond *n.* The leaf of a fern; a large compound leaf of a palm.

Funiculus *n.* The slender basal stalk of an ovule or seed connecting it to the placenta.

Fusain *n.* A brittle, porous type of bituminous coal resembling charcoal, upon fracture it is lustrous black in colour.

Fusinite *n.* A preserved piece of *fusain.*

Gametophyte *n.* The gamete-producing phase in a plant characterized by an alternation of generation.

Gastropod *n.* Any of various molluscs of the class Gastropoda, such as the snail, slug, cowrie, or limpets, characteristically having a single, usually coiled shell or no shell at all, a *ventral* muscular foot for locomotion, and eyes and feelers located on a distinct head.

Gemma *n.*; **Gemmae** *n. pl.* An asexual propagule as in liverworts, capable of developing into a new individual; a bud.

Genus *n.* A taxonomic category ranking below a family and above a species and generally consisting of a group of species exhibiting similar characteristics.

Globose *adj.* Spherical; globular.

Gondwana *n.* The hypothetical protocontinent of the southern hemisphere that, according to the theory of plate tectonics, broke up into India, Australia, Africa, South America, and Antarctica.

Gymnosperm *n.* A plant, such as a cycad or conifer, whose seeds are not enclosed within an ovary.

Haplocheilic *adj.* In *gymnosperms*, a *stomatal* apparatus in which the *subsidiary cells* are not derived from the same initial as the guard cells, as occurs in the cycads.

Hash *n.* A jumble; a hodgepodge.

Helical *adj.* Of or having the shape of a *helix*; a three-dimensional spiral.

Helix *n.* A three-dimensional curve that lies on a cylinder or cone, so that its angle to a plane perpendicular to the *axis* is constant.

Herbaceous *adj.* Relating to or characteristic of an herb as distinguished from a woody plant.

Heterosporous *adj.* Producing two types of spores differing in size and sex.

Hilum n. The scar on a seed, such as a bean, indicating the point of attachment to the funiculus.

Homonym *n.* A taxonomic name identical to one previously applied to a different species or *genus* and therefore unacceptable in its new use.

Homosporous *adj.* Producing spores of one kind only.

Hybridization *n.* The act of mixing different species or varieties of plants to produce a hybrid.

Hyphae *n.* The long branching filaments of which the mycelium (and the greater part of the plant) of a fungus is formed.

Ichnofossil *n.* Fossil trace left by a biological organism, e.g., fossilized footprint, worm burrow.

Imbricate *adj.* Having regularly arranged, overlapping edges.

Impression *n.* A concavity in a surface produced by pressing.

Indehiscent adj. Not splitting open at maturity.

Indusium *n.* An *epidermal* membranous outgrowth covering the sori in some ferns.

Inflorescence *n.* The general arrangement and distribution of flowers on an *axis*.

Infructescence *n.* The fruiting stage of an *inflorescence*.

In situ *abbrev.* **In situation**. Not moved, transported or re-deposited.

integument *n.* The envelope of an ovule.

Interfascicular *n.* Between fascicles or bundles in a stem.

Internode *n.* The region between nodes of a stem.

Interseminal scale *n.* A sterile scale found in the seed cones of the *Bennettitales*.

Inverted *v.* To turn inside out or upside down.

Ironstone *n.* Common name used to describe fossilized mudstones containing iron carbonate, which oxidize red to reddish orange upon exposure to natural weathering.

Lag *n.* **Lag deposit**. A residual accumulation of coarse fragments that remains on the surface after finer material has been removed either by wind or *fluvial* processes, e.g., a lag deposit in the bend of a river.

Lamina *n.*; **laminae** *n. pl.* The thin, flat expanded portion of a leaf blade or petal.

Latewood *n.* The portion of the annual ring that is formed after formation of earlywood has ceased.

Lenticular *adj.* Shaped like a biconvex lens.

Lignification n. The process of turning into wood or becoming woodlike.

Lineage *n.* The descendants of a common ancestor considered to be the founder of the line.

Lithology *n.* The gross physical character of a rock or rock formation.

Locule *n.* A small cavity or compartment within an organ or part of a plant, as any cavities with an ovary or the cavity within a sporangium containing *spore*.

Longitudinally *adj.* Of or relating to longitude or length.

Maastrichtian *n.* The last stage of the Cretaceous period, and therefore of the Mesozoic era. It spanned from 70.6 ± 0.6 Ma to 65.5 ± 0.3 Ma (million years ago).

Marcescent *adj.* Withering but not falling off.

Massa *n.*; **massae** *n. pl.* A specialized structure composed of aborted spores and tapetal material found on certain megaspores of the Azollaceae and often improperly referred to as a floating apparatus or *floats*.

Massula *n.*; **massulae** *n. pl.* A general term for aggregations of pollen grains dispersed as a unit. Within *spore*, massula is defined as the expansion and coalescence of the perine layer to produce the pseudo-cellular material in which multiple microspores are imbedded and dispersed as a unit.

Matrix *n.* Solid matter in which a fossil or crystal is embedded.

Maturation *n.* The process of becoming mature.

Medial *adj.* Relating to, situated in, or extending toward the middle; *median*.

Median *adj.* Relating to, located in, or extending toward the middle.

Megaplant *n.* In fossils, plant parts that are larger than pollen or spore remains, these can range from large sections of *cuticle* to whole plants.

Megasporangium *n.*; **megasporangia** *n. pl.* A structure that produces one or more megaspores.

Megaspore *n.* The larger of two spores that gives rise to a female *gametophyte*.

Meristematic *adj.* relating to, or pertaining to, the undifferentiated plant tissue from which new cells are formed.

Mesarch *adj.* Describing *xylem maturation* in which the older cells (*protoxylem*) are in the centre of the *xylem* strand because *maturation* has progressed both centrifugally and centripetally.

Mesophyll *n.* The photosynthetic tissue of a leaf, located between the upper and lower *epidermis*.

Mesophytic *n.* Being or growing in or adapted to a moderately moist environment.

Metaxylem *n.* The part of the *xylem* that differentiates after the *protoxylem* and is characterized by broader vessels and *tracheids*.

Micropyle *n.* A minute opening in the integument of an *ovule* of a seed plant through which the pollen tube usually enters.

Microsporangium *n.*; **microsporangia** *n. pl.* A structure in which microspores are formed.

Microspore *n.* The smaller of two types of *spore* that give rise to the male *gametophyte*.

Microsporophyll *n.* A leaf-like structure that bears microsporangia.

Miospore *n.* Arbitrarily defined as a *spore* or pollen grain of less than 200 μm in diameter, regardless of biological function.

Monocotyledonae *n.* A class comprising seed plants that produce an *embryo* with a single *cotyledon* and parallel veined leaves.

Monoecious *adj.* Having unisexual reproductive organs or flowers, with the organs or flowers of both sexes borne on a single plant.

Monograph *n.* A scholarly piece of writing of essay or book length on a specific, often limited subject.

Monophyletic *adj.* Relating to, descended from, or derived from a one stock or source.

Monosporous *adj.* Producing a *spore* of one form and size only.

Monotypic *adj.* Consisting of only one type.

Morphology *n.* The form or structure of an organism or one of its parts.

Mould *n.* The distinctive form in which a thing is made.

Mucro *n.* A short, sharp, abrupt tip or terminal point.

Mucronate *adj.* Of or having a mucro; ending abruptly in a sharp point.

Mudstone *n.* A fine-grained, dark grey sedimentary rock, formed of silt and clay and similar to shale but without laminations.

Mybp *abbrev.* Million years before present.

Mycorrhizal fungi *n.* The symbiotic relationship between certain non-pathological or weakly pathologic fungi and the living cells of roots of certain higher plants.

Mycotrophic *adj.* Describing a symbiotic association between a fungus and the whole of the plant. Such an association occurs when a *mycorrhizal fungus* extends into the aerial pats of the plant, as in certain heathers and orchids.

Niche *n.* The function or position of an organism or population within an ecological community.

Node *n.* The point on a stem where the leaf is attached or has been attached; a joint.

Non-porate *adj.* Of pollen grains not having a pore or pores in the exine.

Nucellus *n.* The central part of an ovule, containing the embryo sac.

Oblique *adj.* Having sides of unequal length or form.

Obtuse *adj.* Having a blunt or rounded tip.

Octagonal *adj.* Having eight sides and eight angles.

Omega shape *adj.* Shaped like the last (24th) letter of the Greek alphabet.

Organic trace *n.* The trace remains of complex organic (carbon-based) compounds.

Orientation *n.* The act of orienting or the state of being oriented.

Ovule *n.* A minute structure in seed plants, containing the *embryo* sac and surrounded by the nucellus that develops into a seed after fertilization.

Ovuliferous scale *n.* The scale on which the ovules, of certain *gymnosperms*, are attached.

Paleobotany *n.* The branch of paleontology that deals with plant fossils and ancient vegetation.

Paleoecology *n.* The branch of ecology that deals with the interaction between ancient organisms and their environment.

Paleomagnetic *n.* Referring to magnetic field present in fossilized rocks, created by the earth's magnetic field when the rocks were formed.

Paleomagnetostratigraphy *n. Stratigraphy* of sediments based on palaeomagnetic analysis.

Palisade layer *n.* A layer of vertically elongated *parenchyma* cells, such as are seen beneath the *epidermis* of the upper surface of many leaves.

Palmate *adj.* Having three or more veins, leaflets, or lobes radiating from one point.

Palynology *n.* The scientific study of spores and pollen.

Palynomorph *n.* A fossil *spore*, pollen, or dinoflagellate.

Panicle *n.* A branched cluster of flowers in which the branches are racemes.

Pantropical *adj.* Distributed throughout the tropics.

Papilla, *n.*; papillae *n. pl.* A minute projection on the surface of a stigma, petal, or leaf.

Papillate *adj.* A minute projection on the surface of a *palynomorph*, stigma, petal, or leaf.

Parasitic *adj.* Of, relating to, or characteristic of a parasite.

Parenchyma *n.* The primary tissue of higher plants, composed of thin-walled cells and forming the greater part of leaves, roots, the pulp of fruits and the *pith* of stems.

Parenchymatous *adj.* Of, relating to, made up of, or affecting parenchyma.

Paucity *n.* Smallness in number, fewness.

Peduncle *n.* The stalk of an *inflorescence* or a stalk bearing a single flower in a one-flowered *inflorescence*.

Peltate *adj.* Having a flat circular structure attached to a stalk near the centre; shield-shaped.

Pentagonal *n.* A polygon having five sides and five angles.

Pentarch root *n.* Referring to five *xylem* ridges in the root of a higher plant.

Perennial *adj.* Living three or more years.

Pericycle *n.* A plant tissue characteristic of the roots, located between the *endodermis* and *phloem*.

Perinal *adj.* Originating or belonging to the perine region of a *spore*.

Perinal prolongation *n.* In spores, an enlargement of perine sculpture elements to project beyond the general surface of the perine.

Perine *n.* An outer (perisporial) spore wall, present in some families and genera of ferns.

Peripheral *adj.* Of the surface or outer part of a body or organ; external.

Persistent *adj.* Lasting past maturity without falling off, as the scales of a pine cone.

Petiolate *adj.* Having a *petiole*.

Petiole *n.* The stalk by which a leaf is attached to a stem.

Petrification *n.* A process of *fossilization* in which dissolved minerals replace organic matter.

Phloem *n.* The food conduction tissue in vascular plants, consisting of sieve tubes, fibres, *parenchyma*, and sclereids.

Phosphatization *n.* To reduce to the form or condition of a phosphate.

Photosynthate *n.* A chemical product of photosynthesis.

Phyllotactic *adj.* Of or pertaining to phyllotaxy.

Phyllotaxy *n.* The principals governing leaf arrangement; the arrangement of leaves on a stem.

Phylogeny *n.* The evolutionary development and history of a species or higher taxonomic grouping of organisms.

Phylum *n.* A primary division of a kingdom, as of the plant kingdom, ranking next above a class in size.

Pinna *n.*; **pinnae** *n. pl.* A leaflet or primary division of a pinnately compound leaf.

Pinna/root *n. Pinna* modified into a filamentous underwater structure to function in nutrient uptake or *spore* dispersal. Pinna/root are restricted to the Salviniales.

Pinnate *adj.* Having parts or branches arranged on each side of a common *axis*.

Pinnule *n.* Any of the ultimate leaflets of a bipinnately compound leaf.

Pith *n.* The soft, sponge-like, central cylinder of the stems of most flowering plants; composed mainly of *parenchyma*.

Polymorphic *adj.* Having or occurring in several distinct forms.

Polyphyletic *adj.* Relating to or characterized by development from more than one ancestral type.

Polyploid *n.* Having more than two complete sets of chromosomes. Polyploid plants can be either fertile or sterile. Polyploid plants, if viable, are often larger or more productive than diploid plants, and plant breeders often deliberately produce such plants by crossing species or other means. In the animal kingdom, polyploidy is abnormal and often fatal.

Precursor *n.* One that precedes another; a forerunner or *predecessor*.

Predecessor *n.* An object that precedes another in time.

Procambium *n.* The meristem or growing layer in the tip of a stem or root, which gives rise to primary phloem, primary xylem, and cambium.

Procumbent *adj.* Trailing along the ground but not rooting.

Progenitor *n.* An originator of a line of descent; a *precursor*.

Protoplast *n.* A unit of protoplasm, such as makes up a single cell exclusive of the cell wall.

Protoxylem *n.* The first formed *xylem* that differentiates from the procambium.

Proximal *adj.* Nearer to the point of reference such as an origin, a point of attachment, or the midline of the body.

Pseudo-bijugacy *n.* The false paired phyllotactic systems interpreted in various plant genera, e.g., Torreya is interpreted as bijugate but is actually an 8/21 *phyllotactic* system.

Pseudo-opposite phyllotaxy *n.* The false opposition of organs or leaves on a branch.

Punctuated *v.* To be interrupted periodically.

Quadripinnate *n.* Fourth order of leaflets produced from a *rachis*.

Raceme *n.* A cluster of stalked flowers branching off of a single stem.

Rachis *n. Axis* of a compound leaf or compound *inflorescence*.

Radial section *n.* In xylotomy, a longitudinal cut paralleling the rays in a section of wood.

Radicle *n.* The part of a plant *embryo* that develops into a root.

Radula *n.* A flexible tongue-like organ in certain molluscs, having rows of horny teeth on the surface.

Raphe *n.* The portion of the funiculus that is united to the *ovule* wall commonly visible as a line or ridge on the seed coat.

Refugium *n.*; **refugia** *n. pl.* An area that has escaped ecological changes occurring elsewhere and so provides a suitable habitat for relict species.

Reniform *adj.* shaped like a kidney.

Rhizome *n.* A horizontal, usually underground stem that often sends out roots and shoots from its nodes.

Rip up lag *n.* A *lag* formed from the pulling or ripping up of the underlying plant or animal material from lower horizons in a *fluvial* system usually during flood stages or catastrophic events; e.g., underlying plant *debris* from coal swamp ripped up and redeposited immediately on top in overlying mud.

Rosette *n.* A circular cluster of leaves that radiate from a centre at or close to the ground.

Samara *n.* A dry, indehiscent, winged, often one-seeded fruit, as of the ash, elm or maple.

Sandstone *n.* A sedimentary rock formed by the consolidation and compaction of sand and held together by a natural cement such as calcite or silica.

Saprophytic *adj.* Feeding on dead or decaying organic matter as in some plants and fungi.

Sarcotesta *n.* The usually parenchymatous outer layer of the integument in some seeds.

Scalariform *adj.* Resembling the rungs of a ladder; ladder-like.

Sclereid *n*. A thick-walled lignified plant cell that is often branched.

Sclerenchyma *n*. Strengthening tissue composed of relatively short cells (sclereids) and /or relatively long ones (fibres) with thick, often lignified, cell walls and usually lacking a living protoplast at maturity. *Sclerenchyma* cells usually possess simple unbordered pits, although fibre *tracheids* may have pits with a slightly raised border. *Sclerenchyma* may form by thickening (sclerification) of the secondary cell walls of *parenchyma* cells, often involving lignification, or it may develop directly from meristematic tissue.

Sclerotesta *n*. The middle hard layer of the testa in various seeds.

Sclerotic *adj*. Hardened or thickened. Sclerotic *parenchyma* being applied to tissue composed of cells with the walls hardened but not thickened, and *sclerenchyma* to tissue composed of cells with the walls both hardened and thickened.

Siderite *n*. Iron ore in the form of ferrous carbonate.

Sieve plate *n*. The perforated end wall of a sieve tube cell.

Sieve tube *n*. A tube consisting of an end-to-end series of thin-walled living plant cells characteristic of the phloem and held to function chiefly in translocation of organic solutes.

Sieve-tube (cell) *n*. (in angiosperms) a specialized cell derived from the same parent cell as the closely associated companion cell immediately adjacent to it; sieve-tube cells are elongated cells with sieve plates; sieve-tube cells form sieve tubes through which photosynthate is transported.

Silicification *n*. The process of becoming silicified; conversion into silica.

Silicify *v*. To convert into, impregnate with, silica.

Solenostele *n*. A fern *rhizome stele* in which *phloem* is both inside and outside of the primary *xylem*. Further more two endodermal cylinders are developed, one separating the *cortex* from the external *phloem* and the other situated between the internal *phloem* and the *pith*; this is also called an *amphipholic siphonostele*.

Sorus *n*.; **sori** *n. pl*. A cluster of *sporangia* born on the underside of a fern *frond*.

Speculative *adj*. Of, characterized by, or based upon, contemplative speculation.

Spiculate *adj*. Covered with, or having, spicules.

Spicule *n*. A small needle-like structure or part, such as a silicate or calcium carbonate process.

Sporangia *n*. A single-celled or many-celled structure in which spores are produced, as in fungi, algae, mosses, and ferns.

Spore *n*. A small usually single-celled reproductive body that is highly resistant to desiccation and heat and is capable of growing into a new organism, produced especially by certain bacteria, fungi, algae, and non-flowering plants.

Sporocarp *n*. Specialized leaf branch in certain semi-aquatic ferns that encloses the *sori*.

Sporophyll *n*. A leaf or leaf-like organ that bears spores.

Sporopollenin *n*. Substance similar to suberin and *cutin* but more resistant to decay that is found in the exine of pollen grains.

Steinkern *n*. Rock material formed from consolidated mud or sediment that filled a hollow organic structure, such as a fossil shell. The fossil formed after *dissolution* of the mold. Also known as endocast or internal cast.

Stele *n* The central core of tissue in a stem or root of a vascular plant, consisting of *xylem* and *phloem* together with supporting tissues.

Stipe *n*. A supporting stalk or stem-like structure, especially the stalk of a pistil, the *petiole* of a fern or the stalk that supports the cap of a mushroom.

Stipular wing *n*. Lateral wing-like tissue growth on the lower section of the *stipe* in some ferns.

Stoma *n*.; **stomata** *n. pl*. One of the minute pores in the *epidermis* of a leaf or stem through which gases and water vapour pass.

Stomatal band *n*. A defined linear area in which *stomata* are found.

Stomatal configuration *n*. Arrangement of cells or parts of a *stoma*; *stomata* form relationships to one another.

Stomatal distribution *n*. The spatial array of *stoma* on a leaf, stem or other organ.

Stratigraphy *n*. The study of rock strata, especially the *distribution*, *deposition*, and age of sedimentary rocks.

Strobilus *n*. A cone-like structure, such as a pine cone, the fruit of the hop, or a cone of a club moss that consists of overlapping sporophylls spirally arranged along a central *axis*.

Suberin *n*. A waxy waterproof substance present in the cell walls of cork tissue in plants.

Sub-opposite *adj*. Leaves, branches, or organs produced almost in opposition, not truly opposite.

Subsidiary cells *n*. An *epidermal* cell associated with a *stoma* and morphologically distinct from other *epidermal* cells.

Syndetocheilic *adj*. Describing a *gymnosperm stomatal* complex in which the *subsidiary cells* are derived from the same initials as the guard cells, as occurred in the *Bennettitales*.

Taxodiaceae *n*. Coniferous trees, traditionally considered an independent family, though recently included in Cupressaceae in some classification systems.

Taxon *n*.; **taxa** *n. pl*. A taxonomic category or group, such as a *phylum*, order, family, *genus*, or species.

Taxonomy *n*. The classification or organisms in an ordered system that indicates natural relationships.

Terminal *adj*. Growing or appearing at the end of a stem, branch, stalk, or similar part.

Testa *n*. The hard outer covering or integument of a seed.

Tetrarch *n*. Referring to four *xylem* ridges in the root of a higher plant.

Thallus *n*.; **thalli** *n. pl*. A plant body undifferentiated into stem, leaf, or root.

Torsion *n*. The act of twisting or turning.

Tracheid *n*. A cell in the *xylem* of vascular plants.

Transverse *adj*. Situated or lying across; crosswise.

Triaperaturate *adj*. A modification of the exine in certain pollen to have three evenly spaced equatorial loci for the exit of contents.

Tribe *n*. A number of species or genera having certain structural characteristics in common, e.g., a *tribe* of plants.

Trichome *n*. A hair-like or bristle-like outgrowth, as from the *epidermis* of a plant.

Trilacunar node *n*. Nodal structure of a branch or stem possessing three leaf gaps associated with a single leaf being supplied by three leaf traces.

Tripinnate *adj*. Divided into *pinnae* that are subdivided into smaller, further subdivided leaflets or lobes, as in many ferns.

Triprojectate *n*. A group designation (Triprojectacites) for Aquilapollenites and similar, presumably related, forms of late Cretaceous–early Cenozoic *angiosperm* pollen, in which three colpi are borne on the projected ends of three colpal arms.

Tuber *n*. A swollen, fleshy, underground section of a plant, formed from a root, attached to lateral connectives on both ends and not bearing foliage but bearing buds from which new plant shoots arise.

Umbo *n*. A small protuberance on a plant part, e.g., the small woody knob on the external face of a pine *bract*.

Uniseriate *adj*. Arranged in a single row.

Vascular bundle *n*. A group of vessels forming a bundle.

Vasculature *n*. Arrangement of vessels in a stem, leaf, or other organ.

Vegetative *adj*. Of, relating to, or functioning in processes such as growth or nutrition rather than sexual reproduction.

Venation *n*. *Distribution* or arrangement of a system of veins, as in a leaf *blade*.

Ventral *adj*. Of or on the lower or inner surface of an organ that faces the *axis*; *adaxial*.

Vertical *adj*. Being or situated at right angles to the horizon; upright.

Whorled *adj*. Having or forming an arrangement of three or more leaves, petals, or other organs radiating from a single *node*.

Wither *v*. To dry up or shrivel from, or as if from loss of moisture.

Xylem *n*. The supporting and water-conducting tissues of vascular plants; consisting primarily of *tracheids* and vessels.

Xylotomy *n*. The preparation of sections of wood for microscopic inspection.

Zoophilous *adj*. Pollinated by animals.

REFERENCES
AND SUGGESTED
READINGS

Aase, H. C. 1915. Vascular anatomy of the mega-sporophylls of conifers. *Botanical Gazette* 60:277–313.

Adler, I. 1975. A model of space filling in phyllotaxy. *Journal of Theoretical Biology* 53:435–444.

Alvin, K. L., D. H. Dalby, and F. A. Oladele. 1982. Numerical analysis of cuticular characters in Cupressaceae. In *The Plant Cuticle*, edited by D. F. Cutler et al., 379–396. London: Academic Press (for the Linnean Society of London).

Arnold, C. A. 1953. Origin and relationships of the Cycads. *Phytology* 3:51–65.

Arnold, C. A., and J. S. Lowther. 1955. A new Cretaceous conifer form northern Alaska. *American Journal of Botany* 42:522–528.

Aulenback, K. R., and D. R. Braman. 1991. A chemical extraction technique for the recovery of silicified plant remains from ironstones. *Review of Palaeobotany and Palynology* 70:3–8.

Aulenback, K. R., and B. A. LePage. 1998. *Taxodium wallisii* sp. nov. first occurrence of *Taxodium* from the Upper Cretaceous. *International Journal of Plant Sciences* 159:367–390.

Baikovskaya, T. N. 1956. Upper Cretaceous floras of northern Asia. *Palaeobotanica* 2: 49–181 (in Russian).

Bateman, R. M. 1996. Nonfloral homoplasy and evolutionary scenarios in living and fossil land plants. In *Homoplasy: The Recurrence of Similarity in Evolution*, edited by M. J. Sanderson and L. Hufford, 91–130. San Diego: Academic Press.

Batten, D. J., and A. M. Zavattieri. 1995. Occurrence of dispersed seed cuticles and similar microfossils in mainly Cretaceous successions of the Northern Hemisphere. *Cretaceous Research* 16:73–94.

Bell, A. D. 1976. Computerized vegetative mobility in rhizomatous plants. In *Automata, Languages, Development*, ed. A. Lindenmayer and G. Rozenberg, 3–14. Amsterdam: North-Holland.

———. 1980. The vascular pattern of a rhizomatous Ginger (*Alpinia speciosa* L. Zingiberaceae). – 1. The rhizome. *Annals of Botany* 46:213–220.

Bell, W. A. 1949. Uppermost Cretaceous and Paleocene floras of western Alberta. Geological Survey of Canada Bulletin 13.

———. 1963. Upper Cretaceous floras of the Dunvegan, Bad Heart, and Milk River Formations of Western Canada. *Geological Survey of Canada Bulletin* 94:1–76.

———. 1965. Illustrations of Canadian fossils; Upper Cretaceous and Paleocene plants of western Canada. Geological Survey of Canada Paper 65–35.

Bharadwaj, K., and K. N. Kaul. 1981. *Trapa*: Fossil record, distribution and systematics. *Geophytology* 1:195–203.

Binda, P. L. 1968. New species of Spermatites from the Upper Cretaceous of southern Alberta. *Revue de Micropaléontologie.* 11:137–142.

Binda, P. L., and E.W.V. Nambudiri. 1983. Fossil seed cuticles from the Upper Cretaceous Whitemud beds of Alberta and Saskatchewan, Canada. *Canadian Journal of Botany* 61:2717–2728.

Bocher, T. W. 1964. Morphology of the vegetative body of *Metasequoia glyptostroboides*. *Dansk Botanisk Arkiv* 24:1–70.

Bogner, J., G. L. Hoffman, and K. R. Aulenback. 2005. A fossilized aroid infructescence, *Albertarum pueri* gen. nov. et sp. nov, of late Cretaceous (late Campanian) age from the Horseshoe Canyon Formation of southern Alberta, Canada. *Canadian Journal of Botany* 83:591–598.

Bonde, S. D. 2000. *Rhodospathodendron tomlinsonii* gen et sp. nov., an araceous viny axis from the Nawargaon Intertrappean beds of India. *Palaeobotanist* 49:85–92.

———. 2005. *Eriospermocormus indicus* gen. et sp. nov. (Liliales Eriospermaceae): First record of a monocotyledonous corm from the Deccan Intertrappean beds of India. *Cretaceous Research* 26:197–205.

Bonnet, A.L.M. 1955. Contributions à l'étude des Hydropteridees. I. Recherches sur *Pilularia globulifera* et *P. minuta*. *Cellule (Lierre:1884)* 57:129–237.

Boodle, L. A. 1901. Comparative anatomy of the Hymenophyllaceae, Schizaeaceae and Gleicheniaceae. *Annals of Botany* 15:359–421.

Boulter, M. C., and Z. Kvaček. 1989. The Paleocene flora of the Isle of Mull. Incorporating unpublished observations by A. C. Seward and W. N. Edwards. Special Paper, *Palaeontology* 42: 1–149.

Bower, F. O. 1923. The ferns (Filicales); Vol. 1: Analytical examination of the criteria of comparison. New York: Macmillan.

———. 1926. The ferns (Filicales); Vol. 2: The eusporangiate and other relatively primitive ferns. New York: Macmillan.

———. 1928. The ferns (Filicales); Vol. 3: The leptosporangiate ferns. New York: Macmillan.

Boyd, A. 1992. Revision of the late Cretaceous Pautut Flora from west Greenland: Gymospermopsida (Cycadales, cycadeoidales, Caytoniales, Ginkgoales, coniferales). *Palaeontology Abt. B* 225:105–172.

———. 1998. Bennettitales from the early Cretaceous floras of west Greenland; *Pseudocycas* Nathorst. *Palaeontology Abt. B* 247:123–155.

Braman, D. 1992. Field guide to the geology of the Edmonton Group. Royal Tyrrell Museum internal document.

Bremer, K. 2000. Early Cretaceous lineages of monocot flowering plants. *Evolution, PNAS* 97:4707–4711.

Brinkman, D. B. 2003. A review of nonmarine turtles from the late Cretaceous of Alberta. *Canadian Journal of Earth Sciences* 40:557–571.

Brinkman, D. B., and D. A. Eberth. 2006. Turtles of the Horseshoe Canyon and Scollard Formations – further evidence for a biotic response to late Cretaceous climate change. *Proceedings of the Symposium on Turtle Origins, Evolution and Systematics. Fossil Turtle Research* 1:11–18.

Brown, R. W. 1935. Paleobotany. – Some fossil conifers from Maryland and North Dakota. *Journal of the Washington Academy of Science* 25:441–450.

———. 1937. Fossil plants from the Colgate Member of the Fox Hills Sandstone and adjacent strata. United States Geological Survey Professional Paper 189–I: 239–275.

———. 1962. Paleocene flora of the Rocky Mountains and Great Plains. United States Geological Survey Professional Paper 375.

Brown, W. H. 1935. *The Plant Kingdom: A Textbook of General Botany*. Boston: Athenaeum Press.

Brunsfeld, S. J., P. S. Soltis, D. E. Soltis, P. A. Gadek, C. J. Quinn, D. D. Strenge, and T. A. Ranker. 1994. Phylogenetic relationships among the genera of Taxodiaceae and Cupressaceae: Evidence from rbcl sequences. *Systematic Botany* 19:253–262.

Buchholz, J. T. 1938. Cone formation in *Sequoia gigantea*. I. The relation of stem size and tissue development to cone formation. II. The history of the seed cone. *American Journal of Botany* 25:296–305.

Bůžek, C., M. Konzalová, and Z. Kvaček. 1971. The genus *Salvinia* from the Tertiary of the North-Bohemian Basin. *Sbornik Geologickych Ved. Praha, Rada P. Paleontologie* 13:179–222.

Calvert, H. E., S. K. Perkins, and G. A. Peters. 1983. Sporocarp structure in the heterosporous water fern *Azolla mexicana* Presl. *Scanning Electron Microscopy* 3:1499–1510.

Camefort, H. 1956. Etude de la structure du point végétatif et des variations phyllotaxiques chez quelques Gymnospermes. *Annales Sciences Naturelles, Botanique, Biologie végétale* 17:1–185.

Campbell, D. H. 1893. On the development of *Azolla filiculoides*. *Annals of Botany* 8:155–187.

Catterall, R. A., and S. K. Srivastava. 1985. *Aquilapollenites* tetrads from the Maastrichtian Edmonton Group of Alberta, Canada, and their affinity. *Pollen et Spores* 26:391–412.

Chandler, M.E.J. 1922. *Sequoia couttsiae*, Heer, at Hordle, Hants: A study of the characters which serve to distinguish *Sequoia* from *Athrotaxis*. *Annals of Botany* 36:385–390.

———. 1961. The Lower Tertiary floras of Southern England. I Paleocene floras London Clay flora (supplement). British Museum (Natural History), London.

———. 1962. The Lower Tertiary floras of Southern England. II flora of the Pipe-clay series of Dorset (Lower Bagshot). British Museum (Natural History), London.

Chandrasekharam, A. 1974. Megafossil flora from the Genesee locality, Alberta, Canada. *Palaeontology Abt. B* 147:1–41.

Chaney, R. W. 1950. A revision of fossil *Sequoia* and *Taxodium* in western North America based on the recent discovery of *Metasequoia*. *Transactions of the American Philosophical Society* 40:169–263.

Chaturvedi, S. 1993. Morphological, cuticular and anatomical studies of some members of Taxodiaceae. *Bionature* 13:201–205.

Chitaley, S. D., and S. A. Paradkar. 1972. *Rodeites* Sahni reinvestigated – I. *Botanical Journal of the Linnean Society* 65:109–117.

———. 1973. *Rodeites* Sahni reinvestigated – II. *Palaeobotanist* 20:293–297.

Christophel, D. C. 1976. Fossil floras of the Smoky Tower locality, Alberta, Canada. *Palaeontology Abt. B* 157:1–43.

Chrysler, M. A., and D. S. Johnson. 1939. Spore production in *Regnellidium*. *Bulletin of the Torrey Botanical Club* 66:263–279.

Church, M. A. 1920. On the interpretation of phenomena of phyllotaxis. *Botanical Memoirs* 6:1–58. Oxford: Oxford University Press.

Collinson, M. E. 1991. Diversification of modern heterosporous pteridophytes. In *Pollen and Spores*, edited by S. Blackmore and S. H. Barnes, 119–150. Systematics Association Special Volume No. 44. Oxford: Systematics Association and Oxford University Press.

Conner, W. H., and J. W. Day Jr. 1976. Productivity and composition of a Baldcypress-Water Tupelo site and a Bottomland Hardwood site in a Louisiana swamp. *American Journal of Botany* 63:1354–1364.

Crabtree, D. R., and C. N. Miller Jr. 1989. *Pityostrobus makahensis*, a new species of silicified pinaceous seed cone from the middle Tertiary of Washington. *American Journal of Botany* 76:176–184.

Crafts, A. S. 1943a. Vascular differentiation in the shoot apex of *Sequoia sempervirens*. *American Journal of Botany* 30:110–121.

———. 1943b. Vascular differentiation of the shoot apices of ten coniferous species. *American Journal of Botany* 30:382–393.

Crane, P. R. 1981. Betulaceous leaves and fruits from the British Upper Palaeocene. *Botanical Journal of the Linnean Society* 83:103–136.

———. 1984. A re-evaluation of *Cercidiphyllum*-like plant fossils from the British early Tertiary. *Botanical Journal of the Linnean Society* 89:199–230.

———. 1989. Paleobotanical evidence on the early radiation of nonmagnoliid dicotyledons. *Plant Systematic Evolution* 162:165–191.

Crane, P. R., and S. Lidgard. 1989. Angiosperm diversification and paleolatitudinal gradients in Cretaceous floristic diversity. *Science* 246:675–678.

Crane, P. R., and R. A. Stockey. 1985. Growth and reproductive biology of *Joffrea speirsii* gen et sp. nov., a *Cercidiphyllum*-like plant from the late Paleocene of Alberta, Canada. *Canadian Journal of Botany* 63:340–364.

Crane, P. R., S. R. Manchester, and D. L. Dilcher. 1990. A preliminary survey of fossil leaves and well-preserved reproductive structures from the Sentinel Butte Formation (Paleocene) near Almont, North Dakota. *Fieldiana, Geo.* New Series No. 20.

Crane, P. R., E. M. Friis, and K. R. Pedersen. 1994. Palaeobotanical evidence on the early radiation of Magnoliid Angiosperms. *Plant Systematic Evolution* [Suppl.] 8:51–72.

Crane, P. R., E. M. Friis, and K. R. Pedersen. 1995. The origin and early diversification of Angiosperms. *Nature* 374:27–33.

Crane, P. R., P. Herendeen, and E. M. Friis. 2004. Fossils and plant phylogeny. *American Journal of Botany* 91:1683–1699.

Crepet, W. L., and T. Delevoryas. 1972. Investigations of North American Cycadeoids: The reproductive biology of Cycadeoidea. *Palaeontology Abt. B* 148:144–169.

Crepet, W. L., and K. C. Nixon. 1998. Fossil Clusiaceae from the late Cretaceous (Turonian) of New Jersey and implications regarding the history of bee pollination. *American Journal of Botany* 85:1122–1133.

Crepet, W. L., K. C. Nixon, and M. A. Gandolfo. 2004. Fossil evidence and phylogeny: The age of major Angiosperm clades based on mesofossil and macrofossil evidence from Cretaceous deposits. *American Journal of Botany* 91:1666–1682.

Cross, G. L. 1940. Development of the foliage leaves of *Taxodium distichum*. *American Journal of Botany* 27:471–482.

———. 1941. Some histogenetic features of the shoot of *Cryptomeria japonica*. *American Journal of Botany* 28:573–582.

———. 1942. Structure of the apical meristem and development of the foliage leaves of *Cunninghamia lanceolata*. *American Journal of Botany* 29:288–301.

———. 1943a. A comparison of the shoot apices of the Sequoias. *American Journal of Botany* 30:130–142.

———. 1943b. The shoot apices of *Athrotaxis* and *Taiwania*. *Bulletin of the Torrey Botanical Club* 70:335–348.

Croxdale, J. G. 1978. *Salvinia* leaves. I. Origin an early differentiation of floating and submerged leaves. *Canadian Journal of Botany* 56:1982–1991.

———. 1979. *Salvinia* leaves. II. Morphogenesis of the floating leaf. *Canadian Journal of Botany* 57:1951–1959.

———. 1981. *Salvinia* leaves. III. Morphogenesis of the submerged leaf. *Canadian Journal of Botany* 59:2065–2072.

Daghlian, C. P. 1981. A review of the fossil record of Monocotyledons. *Botanical Review* 47:517–555.

Dahlgren, R.M.T., H. T. Clifford, and P. F. Yeo. 1985. *The Families of the Monocotyledons: Structure, Evolution and Taxonomy.* Berlin: Springer.

Daugherty, L. H. 1960. *Itopsiderma*. A new genus of the Osmundaceae from the Triassic of Arizona. *American Journal of Botany* 47:771–777.

Dawson, F. M., C. G. Evans, R. Marsh, and R. Richardson. 1994. Uppermost Cretaceous and Tertiary strata of the western Canada sedimentary basin. In *Geological Atlas of the Western Canada Sedimentary Basin*, compiled by G. Mossop and I. Shetsen, 387–406. Canadian Society of Petroleum Geologists and Alberta Research Council, Calgary, Alberta.

De la Sorta, E. R. 1963. Contributions to the knowledge of the neotropical Salviniaceae. *Darwiniana* 12–13:456–520; 612–623.

Delevoryas, T. 1960. Investigations of North American Cycadeoids: Trunks from Wyoming. *American Journal of Botany* 47:778–786.

———. 1962. Two petrified angiosperms from the Upper Cretaceous of South Dakota. *Journal of Palaeontology* 38:584–586.

———. 1963. Investigations of North American Cycadeoids: Cones of Cycadeoidea. *American Journal of Botany* 50:45–52.

———. 1968. Investigations of North American Cycadeoids: Structure, ontogeny and phylogenetic considerations of cones of Cycadeoidea. *Palaeontology Abt. B* 121:122–133.

Deshpande, J. V. 1943. A study of the sporophyte of *Salvinia cucullata* Roxb. *Journal of the Indian Botanical Society* 22:59–84.

Dettmann, M. E., R. E. Molnar, J. G. Douglas, D. Burger, C. Fielding, H. T. Clifford, J. Francis, P. Jell, T. Rich, M. Wade, P. V. Rich, N. Pledge, A. Kemp, and A. Rozefelds. 1992. Australian Cretaceous terrestrial faunas and floras: Biostratigraphic and biogeographic implications. *Cretaceous Research* 13:207–262.

Dilcher, D. L. 1974. Approaches to the identification of Angiosperm leaf remains. *Botanical Review* 40:1–157.

Doi, T., and K. Morikawa. 1929. An anatomical study of the leaves of the genus *Pinus*. J. *Department of Agriculture, Kyushu Imperial University* 2:149–198.

Dolph, G. E., and D. L. Dilcher. 1979. Foliar physiognomy as an aid in determining paleoclimate. *Palaeontographica, Abt. B* 170:151–172.

Doludenko, M. P., and Y. I. Kostina. 1987. On the conifers of the genus *Elatides*. *Palaeontology Journal* 21:120–125.

Dorf, E. 1942. *Upper Cretaceous Floras of the Rocky Mountain Region*. Contributions to Paleontology. Carnegie Institution of Washington Publication 508.

Drinnan, A. N., P. R. Crane, E. M. Friis, and K. R. Pedersen. 1990. Lauraceous flowers from the Potomac Group (Mid-Cretaceous) of eastern North America. *Botanical Gazette* 151:370–384.

———. 1991. Angiosperm flowers and tricolpate pollen of buxaceous affinity from the Potomac Group (mid-Cretaceous) of eastern North America. *American Journal of Botany* 78:153–176.

Drinnan, A. N., P. R. Crane, and S. B. Hoot. 1994. Patterns of floral evolution in the early diversification of non-magnoliid Dicotyledons (Eudicots). *Plant Systematic Evolution [Suppl.]* 8:93–122.

Eberth, D. A. 1992. Tectonic, stratigraphic, and sedimentological significance of a regional discontinuity in the Upper Judith River Group (Belly River wedge) of southern Alberta, Saskatchewan, and northern Montana. *Canadian Journal of Earth Sciences* 30:174–200.

———. 2002. Review and comparison of the Belly River Group and Edmonton Group stratigraphic architecture in the southern Alberta Plains. In *Programs and Abstracts*, Canadian Society of Petroleum Geologists, Diamond Jubilee Convention, Calgary, Alberta, 3–7 June 2002. Canadian Society of Petroleum Geologists, Calgary, p. 117, and Extended Abstracts PDF file, pp. 1–7.

———. 2004. A revised stratigraphy for the Edmonton Group (Upper Cretaceous) and its potential sandstone reservoirs. Field Trip #7. CSPG-CHOA-CWLS Joint Conference, 31 May – 4 June 2004.

Eckenwalder, J. E. 1976. Re-evaluation of the Cupressaceae and Taxodiaceae: A proposed merger. *Madrono* 23:237–300.

Eklund, H. 2000. Lauraceous flowers from the late Cretaceous of North Carolina, U.S.A. *Botanical Journal of the Linnean Society* 132:397–428.

Ellis, C. H., and R. H. Tschudy. 1964. The Cretaceous megaspore genus *Arcellites* Miner. *Micropaleontology* 10:73–79.

Endlicher, S. L. 1847. *Synopsis Coniferarum.* Scheitlin und Zollikofer, Sangalli (Sankt Gallen).

Erdtman, G. 1972. *Pollen and spore morphology/plant taxonomy: Gymnospermae, Pteridophyta, Bryophyta* (illustrations). New York: Hafner.

———. 1986. *Pollen Morphology and Plant Taxonomy: Angiosperms* (An Introduction to Palynology). Leiden: Brill.

Fensome, R. A. 1987. Taxonomy and biostratigraphy of Schizaealean spores from the Jurassic-Cretaceous boundary beds of the Aklavik Range, District of Mackenzie. *Palaeontographica Canadiana* 4:1–49.

Ferguson, D. K. 1978. Some current research on fossil and recent taxads. *Review of Palaeobotany and Palynology* 26:213–226.

Ferre, Y., and M. H. Gaussen. 1968. Systematique vegetale – Les Cupressacees australes. *Comptes-rendus de l'Académie des Sciences de Paris* 267:483–487.

Florin, R. 1952. On two conifers from the Jurassic of south-eastern Australia. *Palaeobotanist* 1:177–182.

———. 1958. On Jurassic taxads and conifers from northwestern Europe and eastern Greenland. *Acta Horti Bergiani* 17:257–402.

Florin, R., and J. B. Boutelje. 1954. External morphology and epidermal structure of leaves in the genus *Libocedrus*, s. lat. *Acta Horti Bergiani* 17:7–37.

Foster, A. S., and E. M. Gifford Jr. 1974. *Comparative Morphology of Vascular Plants.* San Francisco: W.H. Freeman.

Fowler, K., and J. Stennett-Wilson. 1978. Sporoderm architecture in modern *Azolla. Fern Gazette* 11:405–412.

French, J. C., and P. B. Tomlinson. 1983. Vascular patterns in stems of Araceae: Subfamilies Colocasioideae, Aroideae and Pistioideae. *American Journal of Botany* 70:756–771.

———. 1986. Compound vascular bundles in monocotyledonous stems: Construction and significance. *Kew Bulletin* 41:561–574.

Friis, E. M. 1988. *Spirematospermum chandlerae* sp. nov., an extinct species of Zingiberaceae from the North American Cretaceous. *Tertiary Research* 9:7–12.

Friis, E. M., and A. Skarby. 1981. Structurally preserved angiosperm flowers from the Upper Cretaceous of southern Sweden. *Nature* 291:484–486.

Friis, E. M., H. Eklund, K. R. Pedersen, and P. R. Crane. 1994a. *Virginianthus calycanthoides* gen. et sp. nov. – A calycanthaceous flower from the Potomac Group (early Cretaceous) of eastern North America. *International Journal of Plant Sciences* 155:772–785.

Friis, E. M., K. R. Pedersen, and P. R. Crane. 1994b. Angiosperm floral structures from the early Cretaceous of Portugal. *Plant Systematic Evolution [Suppl.]* 8:31–49.

———. 1995. *Appomattoxia ancistrophora* gen. et sp. nov. – A early Cretaceous plant with similarities to *Circaeaster* and living Magnoliidae. *American Journal of Botany* 82:933–943.

———. 2000. Reproductive structure and organization of basal angiosperms from the early Cretaceous (Barremian or Aptian) of western Portugal. *International Journal of Plant Sciences [suppl.]* 161:169–S182.

———. 2004. Araceae from the early Cretaceous of Portugal: Evidence on the emergence of monocots. *Proceedings of the National Academy of Science, U.S.A.* 101:16565–16570.

Fulling, E. H. 1934. Identification, by leaf structure, of the species of *Abies* cultivated in the United States. *Bulletin of the Torrey Botanical Club* 61:497–524.

Gandolfo, M., K. C. Nixon, W. L. Crepet, and G. E. Ratcliffe. 1997. A new fossil fern assignable to Gleicheniaceae from late Cretaceous sediments of New Jersey. *American Journal of Botany* 84:483–493.

Gandolfo, M., K. C. Nixon, and W. L. Crepet. 1998a. A new fossil flower from the Turonian of New Jersey: *Dressiantha bicarpellata* gen. et sp. nov. (Capparales). *American Journal of Botany* 85:964–974.

———. 1998b. *Tylerianthus crossmanensis* gen. et sp. nov. (aff. Hydrangeaceae) from the Upper Cretaceous of New Jersey. *American Journal of Botany* 85:376–386.

Gandolfo, M. A., D. W. Stevenson, and E. M. Friis. 1998c. Oldest known fossils of monocotyledons. *Nature* 394:532–533.

Gandolfo, M. A., K. C. Nixon, and W. L. Crepet. 2000. Monocotyledons: A review of their early Cretaceous record. *Proceedings of the Second International Conference on the Comparative Biology of the Monocotyledons*, edited by K. Wilson and D. Morrison, 44–52. Sydney, Australia.

———. 2002. Triuridaceae fossil flowers from the Upper Cretaceous of New Jersey. *American Journal of Botany* 89:1940–1957.

Gibson, D. W. 1977. Upper Cretaceous and Tertiary coal-bearing strata in the Drumheller-Ardley region, Red Deer River valley, Alberta. Geological Survey of Canada Paper 76–35.

Golovneva, L. B. 1988. A new genus *Microconium* (Cupressaceae) from the late Cretaceous deposits of the north-east of the USSR. *Botanicheskiy Zhurnal (Leningrad)* 73:1179–1183.

Gregor, H. J., and J. Bogner. 1984. Fossile Araceen Mitteleuropas und ihre rezenten Vergleichsformen. *Documenta Naturae* 19:1–12.

———. 1989. Neue Untersuchungen an Tertiären Araceen II. *Documenta Naturae* 49:12–22.

Greguss, P. 1955. *Identification of Living Gymnosperms on the Basis of Xylotomy.* Budapest: Akadémiai Kiadó.

———. 1956. The phyllotaxy of *Metasequoia*, *Sequoia* and *Taxodium*. *Acta Biologica Szegediensis* 2:29–38.

———. 1966. The relationships of Cycadales on the basis of their xylotomy, branching and leaf epidermis. *Palaeobotanist* 14:94–101.

Gupta, K. M. 1957. Some American species of *Marsilea* with special reference to their epidermal and soral characters. *Madrono* 14:113–127.

———. 1962. *Marsilea.* Botanical Monographs. *New Dehli* 2:1–109.

Hall, J. W. 1969. Studies on fossil *Azolla*: Primitive types of megaspores and massulae from the Cretaceous. *American Journal of Botany* 56:1173–1180.

Hall, T. F., and W. T. Penfound. 1943. Cypress-gum communities in the Blue Girth Swamp near Selma, Alabama. *Ecology* 24:208–217.

Halle, N. 1979. Architecture du rhizome chez quelques Zingiberacees d'Afrique et d'Océanie. *Adansonia* 19:127–144.

Hamblin, A.P. 2004. The Horseshoe Canyon Formation in southern Alberta: Surface and subsurface stratigraphic architecture, sedimentology, and resource potential. Geological Survey of Canada Bulletin 578.

Harris, T. M. 1941. Cones of extinct Cycadales from the Jurassic rocks of Yorkshire. *Philosophical Transactions of the Royal Society of London; Series B* 231:75–98.

———. 1943. The fossil conifer *Elatides williamsoni*. *Annals of Botany N.S.* 7:325–339.

———. 1944. A revision of *Williamsoniella*. *Philosophical Transactions of the Royal Society of London; Series B* 231:313–328.

———. 1953. Conifers of the Taxodiaceae from the Wealden Formation of Belgium. *Institut Royal des Sciences naturelles de Belgique Memoir* 126:1–43.

———. 1961. The fossil Cycads. *Palaeontology* 4:313–323.

———. 1964. The Yorkshire Jurassic flora. II. Caytoniales, Cycadales and Pteridosperms. British Museum (Natural History), London.

———. 1969. The Yorkshire Jurassic flora. III. Bennettitales: Genus *Anomozamites*. British Museum (Natural History), London.

———. 1976. The Mesozoic Gymnosperms. *Review of Palaeobotany and Palynology* 21:119–134.

———. 1979. The Yorkshire Jurassic flora. V. Coniferales. British Museum (Natural History), London.

Hart, J. A. 1987. A cladistic analysis of conifers: Preliminary results. *Journal of the Arnold Arboretum* 68:269–307.

Hasebe, M., P. G. Wolf, K. M. Pryer, K. Ueda, M. Ito, R. Sano, G. J. Gastony, J. Yokoyama, J. R. Manhart, N. Murakami, E. H. Crane, C. H. Haufler, and W. D. Hauk. 1995. Fern phylogeny based on rbcl nucleotide sequences. *American Fern Journal* 85:134–181.

Hayata, B., and Y. Satake. 1929. Contributions to the knowledge of the systematic anatomy on some Japanese plants. *Botanical Magazine* 43:73–106.

Henry, A., and M. McIntyre. 1926. The swamp cypresses, *Glyptostrobus* of China and *Taxodium* of America, with notes on allied genera. *Proceedings of the Royal Irish Academy* 37B 13:90–116.

Herendeen, P. S. 1991. Lauraceous wood from the mid-Cretaceous Potomac group of eastern North America: *Paraphyllanthoxylon marylandense* sp. nov. *Review of Palaeobotany and Palynology* 69:277–290.

Herendeen, P. S., and P. R. Crane. 1995. The fossil history of the monocotyledons. In *Monocotyledons: Systematics and Evolution*, edited by P. J. Rudall, P. J. Cribb, D. F. Cutler, and C. J. Humphries, 1–21. Royal Botanical Gardens Kew.

Herendeen, P. S., D. H. Les, and D.L. Dilcher. 1990. Fossil *Ceratophyllum* (Ceratophyllaceae) form the Tertiary of North American. *American Journal of Botany* 77:7–16.

Herendeen, P. S., P. R. Crane, and A. N. Drinnan. 1995. Fagaceous flowers, fruits, and cupules from the Campanian (late Cretaceous) of central Georgia, USA. *International Journal of Plant Sciences* 156:93–116.

Hernández-Castillo, G., and S.R.S. Cevallos-Ferriz. 1999. Reproductive and vegetative organs with affinities to Haloragaceae from the Upper Cretaceous Huepac Chert Locality of Sonora, Mexico. *American Journal of Botany* 86:1717–1734.

Herzog, R. 1934. Anatomische und experimentell-morphologische Untersuchungen über die Gattung *Salvinia. Planta* 22:490–514.

Hewitson, W. 1962. Comparative morphology of the Osmundaceae. *Annals of the Missouri Botanical Garden* 49:57–93.

Hickey, L. J., and R. K. Peterson. 1978. *Zingiberopsis*, a fossil genus of the ginger family from the late Cretaceous to early Eocene sediments of western interior North America. *Canadian Journal of Botany* 56:1136–1152.

Hida, M. 1957. The comparative study of Taxodiaceae from the standpoint of the development of the cone scale. *Botanical Magazine Tokyo* 70:44–51.

———. 1962. The systematic position of *Metasequoia. Botanical Magazine Tokyo* 75:316–323.

Hills, L. V., and N. Weiner. 1965. *Azolla geneseana*, n. sp., and revision of *Azolla primaeva. Micropaleontology* 11:255–261.

Ho, R. H., and O. Sziklai. 1973. Fine structure of the pollen surface of some Taxodiaceae and Cupressaceae species. *Review of Palaeobotany and Palynology* 15:17–26.

Hoffman, G. L. 2002. Paleobotany and paleoecology of the Joffre Bridge Roadcut Locality (Paleocene), Red Deer, Alberta. PhD thesis, University of Alberta.

Hoffman, G. L., and R. A. Stockey. 1994. Sporophytes, megaspores and massulae of *Azolla stanleyii* from the Paleocene Joffre Bridge Locality, Alberta. *Canadian Journal of Botany* 72:301–308.

Hollick, A. 1930. The Upper Cretaceous floras of Alaska. United States Geological Survey Professional Paper 159.

Hollick, A., and E. C. Jeffrey. 1906. Affinities of certain Cretaceous plant remains commonly referred to the genera *Dammara* and *Brachyphyllum. American Naturalist* 40:189–216.

Hollick, A., and E. C. Jeffrey. 1909. Studies of Cretaceous coniferous remains from Kreischerville, New York. *Memoirs of the New York Botanical Garden* 3:1–76.

Holttum, R. E. 1955. Growth habits of monocotyledons – variations on a theme. *Phytomorphology* 5:399–413.

Hu, H. H., and R. C. Ching. 1927. Synoptical study of Chinese Torreyas with supplemental notes on the distribution and habitat. *Contributions to the Biological Laboratory Sciences Society of China* 3:1–37.

Hu, S. Y. 1951. Notes on the flora of China. *Journal of the Arnold Arboretum* 32:390–402.

———. 1980. The *Metasequoia* flora and its phytogeographic significance. *Journal of the Arnold Arboretum* 61:41–94.

Hughes, N. F. 1994. The enigma of angiosperm origins. *Cambridge Paleobiology*, Series 1. Cambridge: Cambridge University Press.

Jahrig, M. 1962. Beitrage zur Nadelanatomie und Taxonomie der Gattung *Pinus* L. *Willdenowia* 3:329–366.

Jarzen, D. M. 1977. *Aquilapollenites* and some santalalean genera. *Grana* 16:29–39.

———. 1982. Palynology of Dinosaur Provincial Park (Campanian) Alberta. National Museum of Natural Sciences. *Syllogeus* 38:1–69.

Jean, R. V. 1989. Phyllotaxis: A reappraisal. *Canadian Journal of Botany* 67:3103–3107.

———. 1994. *Phyllotaxis*. Cambridge University Press.

Jeffrey, E. C. 1908. On the structure of the leaf in Cretaceous Pines. *Annals of Botany* 22:207–220.

Johnson, D. M. 1986. Systematics of the New World species of *Marsilea* (Marsileaceae). *American Society of Plant Taxonomists, Systematic Botany Monograph* 11:1–87.

Johnson, D. S. 1898. On the leaf and sporocarp of *Pilularia*. *Botanical Gazette* 24:1–24.

———. 1933. Structure and development of *Pilularia minuta. Durieu manuscript. Botanical Gazette* 95:104–127.

Johnson, D. S., and M. A. Chrysler. 1938. Structure and development of *Regnellidium diphyllum*. *American Journal of Botany* 25:141–156.

Jones, D. L. 1987. *Encyclopedia of Ferns*. Portland, OR: Timber Press.

Jongmans, W. J., and S. J. Dijkstra. 1972. Fossilium Catalogus II: Plantae. 79–87. Gymnospermae, *Uitgeverij Dr. W. Junk N. V., 's-Gravenhage*.

Kar, R. K., and D. L. Dilcher. 2002. An argument for the origins of heterospory in aquatic environments. *Palaeobotanist* 51:1–11.

Keating, R. C. 1968. Trends of specialization in the stipe anatomy of *Dennstaedtia* and related genera. *American Fern Journal* 58:126–140.

Kedves, M. 1985. LM, TEM and SEM investigations on recent inaperturate Gymnospermatophyta pollen grains. *Acta Biol. Szeged* 31:129–146.

Kempf, E. K. 1971. Elektronenmikroskopie der sporodermis von mega- und mikrosporen der pteridophyten-Gattung *Salvinia* aus dem Tertiar und Quartar Deutschlands. *Palaeontology Abt. B* 136:47–70.

Kendall, M. W. 1947. XXI. – On five species of *Brachyphyllum* form the Jurassic of Yorkshire and Wiltshire. *Annals and Magazine of Natural History, Ser. 11.* 14:225–251.

———. 1949. XXIV. – On *Brachyphyllum expansum* (Sternberg) Seward, and its cone. *Annals and Magazine of Natural History, Ser. 12.* 2:308–320.

Kerp, H. 1990. The study of fossil Gymnosperms by means of cuticular analysis. *Palaios* 5:548–569.

Khoshoo, T. N. 1959. Polyploidy in Gymnosperms. *Evolution* 13:24–39.

Kimura, T., and S. Sekido. 1975. *Nilssoniocladus* n. gen. (Nilssoniaceae n. fam.), newly found from the early Lower Cretaceous of Japan. *Palaeontology Abt. B* 153:111–118.

Knobloch, E. 1979. *Zingiberopsis riggauensis* sp. n. – an interesting monocotyledon from the Cretaceous of Bavaria. *Vestnik Ustredniho ustavu Geo.* 54:297–300.

Knobloch, E., and D. Mai. 1983. Carbonized seeds and fruits from the Cretaceous of Bohemia and Moravia and their stratigraphical significance. *Mis. Micropalaeo.: A Mem. Vol./ 18th Euro. Coll. Micropaleontology* 4:305–332.

Knowlton, F. H. 1905. Fossil plants of the Judith River beds. *United States Geological Survey Bulletin* 257:129–172.

Kodrul, T. M., M. V. Tekleva, and V. A. Krassilov. 2006. A new conifer species, *Mesocyparis rosanovii* sp. nov. (Cupressaceae, Coniferales), and transberingian floristic connections. *Paleontological Journal* 40:328–338.

Konar, R. N., and R. K. Kapoor. 1972. Anatomical studies on *Azolla pinnata*. *Phytomorphology* 22:211–223.

Kondinskaya, L. I. 1966. Fossil spore of the water ferns in Upper Cretaceous and Paleocene deposits of the west Siberian lowlands. In *Palynology of Siberia*, edited by A. F. Chlonova, 116–122. Academy of Sciences. USSR. Nauka, Moscow. (in Russian).

Kovach, W. L., and D. J. Batten. 1989. Worldwide stratigraphic occurrences of Mesozoic and Tertiary megaspores. *Palynology* 13: 247–277.

Krassilov, V. A. 1976. *Tsagayanskaya flora Amurskoi oblasti* (Tsaganyansk flora of the Amur region). Akademia Nauk SSSR Dalinevostochnoii Nauchnii Shentr Biologo-Pochvennii Institut. Vladivostok: Izdatelstvo Nauka.

Krussmann, G. 1986. *Manual of Cultivated Broad-Leaved Trees and Shrubs*. Vols. I, II, III. Portland, OR: Timber Press.

———. 1991. *Manual of Cultivated Conifers*. Portland, OR: Timber Press.

Kulkarni, A. R., and K. S. Patil. 1975. *Palmocaulon costapalmatum*, a petrified palm leaf axis from the Deccan intertrappean beds of Wardha district, Maharashtra. *Geophytology* 7:208–213.

Kumazawa, M. 1972. Periodic variations of the divergence angle, internode length and leaf shape, revealed by correlogram analysis. *Phytomorphology* 21:376–389.

Kurmann, M. H. 1992. Exine stratification in living Gymnosperms: A review of published transmission electron micrographs. *Kew Bulletin* 47:25–39.

Kusumi, J., Y. Tsumura, H. Yoshimaru, and H. Tachida. 2000. Phylogenetic relationships in Taxodiaceae and Cupressaceae sensu stricto based on *MATK* gene, *CHLL* gene, *TRNL-TRNF* IGS region, and TRNL intron sequences. *American Journal of Botany* 87:1480–1488.

Kutluk, H. 1985. Megaspore palynology of the Bearpaw-Horseshoe Canyon Formation transition zone, Alberta. Master's thesis University of Alberta.

Kvaček, J. 1992. *Nilsonia* and *Nilsonia*-like leaves from the Cenomanian of Bohemia. *Acta Universitatis Carolinae – Geologica* 1–2:63–71.

———. 1995. Cycadales and Bennettitales leaf compressions of the Bohemian Cenomanian, Central Europe. *Review of Palaeobotany and Palynology* 84:389–412.

———. 1997. *Microzamia gibba* (Reuss) Corda: A cycad ovulate cone from the Bohemian Cretaceous basin, Czech Republic- micromorphology and a reinterpretation of its affinities. *Review of Palaeobotany and Palynology* 96:81–97.

Kvaček, J., and A. B. Herman. 2004. Monocotyledons from the early Campanian (Cretaceous) of Grünbach, lower Austria. *Review of Palaeobotany and Palynology* 128:323–353.

Kvaček, Z. 1995. *Limnobiophyllum* Krassilov – a fossil link between the Araceae and the Lemnaceae. *Aquatic Botany* 50:49–61.

Kvaček, Z. 2003. Aquatic angiosperms of the Most Formation of North Bohemia (Central Europe). *Courier Forsch.-Inst. Senckenberg* 241:255–279.

LaPasha, C. A., and C. N. Miller Jr. 1981. New Taxodiaceous seed cones from the Upper Cretaceous of New Jersey. *American Journal of Botany* 68:1374–1382.

Larson, P. R. 1977. Phyllotactic transitions in the vascular system of *Populus deltoids*

Bartr. as determined by C14 labeling. *Planta* 134:241–249.

———. 1983. Primary vascularization and the sitting of primordial. In *The Growth and Functioning of Leaves,* edited by J E. Dale and F. L. Milthorpe, 25–51. Cambridge: Cambridge University Press.

Lebedev, E. L. 1976. Evolution of Albian-Cenomanian floras of northeast USSR and the association between their composition and facies conditions. *International Geology Review* 19:1183–1190.

———. 1979. The paleobotanical basis for the stratigraphy of the volcanogenic rocks in the Ul'inskiy Basin (Okhotsk-Chukotka volcanigenic belt). *Izvestia Akademia Nauka SSSR, Seria Geologia* 10:25–39 (in Russian).

———. 1982. Recurrent development of floras of the Okhotsk-Chukotka volcanogenic belt at the boundary between the Early and Late Cretaceous. *Paleontological Journal* 2:1–11.

———. 1987. Stratigraphy and age of the Okhotsk-Chukotsk volcanogenic belt. *Academy of Sciences of the USSR, Transactions* 421:1–175 (in Russian).

Lehman, T. M. 1987. Late Maastrichtian paleoenvironments and dinosaur biogeography in the western interior of North America. *Palaeogeography, Palaeoclimatology, Palaeoecology* 60:189–217.

Lehman, T. M., and E. A. Wheeler. 2001. A fossil dicotyledonous woodland/forest from the Upper Cretaceous of Big Bend National Park, Texas. *Palaios* 16:102–108.

Lemoine-Sebastian, C. 1968. La vascularisation du complexe bractee-ecaille chez les Taxodiacees. *Travaux du Laboratoire Forestier de Toulouse,* Tome 1, 7:1–22.

———. 1971. Vascularisation du complexe bractee-ecaille dans le cone femelle des Cupressacees. Nouvelles observations. *Botanica Rhedonica, Series A* 11:177–187.

LePage, B. A. 2003. A new species of *Thuja* (Cupressaceae) from the late Cretaceous of Alaska: Implications of being evergreen in a polar environment. *American Journal of Botany* 90:167–174.

LePage, B. A., and J. F. Basinger. 1993. The use of lacquer (Nitrocellulose) for the coating and preservation of fossil leaf compressions. *Journal of Palaeontology* 67:128–134.

LePage, B. A., C. J. Williams, and H. Yang. 2005. The geobiology and ecology of *Metasequoia.* Dordrecht, Netherlands: Springer.

Lerbekmo, J. F., and D. R. Braman. 2002. Magnetostratigraphic and biostratigraphic correlation of late Campanian and Maastrichtian marine and continental strata from the Red Deer Valley to the Cypress Hills, Alberta, Canada. *Canadian Journal of Earth Sciences* 39:539–557.

Lerbekmo, J. F., and K. C. Coulter. 1984. Late Cretaceous to early Tertiary magnetostratigraphy of a continental sequence: Red Deer Valley, Alberta, Canada. *Canadian Journal of Earth Sciences* 22:567–583.

Les, D. H. 1989. The evolution of achene morphology in *Ceratophyllum* (Ceratophyllaceae), IV. Summary of proposed relationships and evolutionary trends. *Systematic Botany* 14:254–262.

Lesquereux, L. 1878. Contributions to the flora of the western territories II. The Tertiary flora. *Report of the United States Geological Survey of the Territories* 7:1–366.

Li, H. 1964. *Metasequoia,* a living fossil. *American Scientist* 52:93–109.

Lidgard, S., and P. R. Crane. 1990. Angiosperm diversification and Cretaceous floristic trends: A comparison of palynofloras and leaf macrofloras. *Paleobiology* 16:77–93.

Liu, T., and H. Su. 1983. Biosystematic studies on *Taiwania* and numerical evaluations of the systematics of Taxodiaceae. *Taiwan Special Publication; Ser.* 2:1–113.

Liu, Y., C. Li, and Y. Wang. 1999. Studies on fossil *Metasequoia* from north-east China and their taxonomic implications. *Botanical Journal of the Linnean Society* 130:267–297.

Loyal, D. S., and R. K. Grewal. 1967. Some observations on the morphology and anatomy of *Salvinia*, with particular reference to *S. auriculata* and *S. natans* All. Research Bulletin of the Panjab University Science 18:13–28.

Lupia, R., H. Schneider, G. M. Moeser, K. M. Pryer, and P. R. Crane. 2000. Marsileaceae sporocarps and spores from the late Cretaceous of Georgia, U.S.A. *International Journal of Plant Sciences* 161:975–988.

Mahabale, T. S. 1954. The genus *Salvinia* and evolutionary problems related to it. *Congres International de Botanique* 7:304–306.

———. 1957. Trends of specialization in the sporocarp and spores in the living and fossil Marsileaceae. *Palaeobotanist* 5:66–72.

Mai Von, D. H. 1970. Subtropische Elemente im europaischen Tertiar I. *Palaeontologisches Abhandlungen* 3:441–503.

———. 1987. Neue Fruchte und Samen aus palaozanen ablagerungen Mitteleuropas. *Feddes Repertorium* 98:197–229.

Mamay, S. H. 1976. Paleozoic origin of the Cycads. *United States Geological Survey Professional Paper* 934:1–46.

Manchester, S. R. 1987. Extinct ulmaceous fruits from the Tertiary of Europe and western North America. *Review of Paleobotany and Palynology* 52:119–129.

———. 1994. Fruits and seeds of the middle Eocene Nut Beds Flora, Clarno Formation, Oregon. *Palaeontology Americana* 58:1–205.

———. 1999. Biogeographical relationships of North American tertiary floras. *Annals of the Missouri Botanical Garden* 86:472–522.

———. 2002. Leaves and fruits of *Davidia* (Cornales) from the Paleocene of North America. *Systematic Botany* 27:368–382.

Manchester, S. R., and Chen Zhi-duan. 1996. *Palaeocarpinus aspinosa* sp. nov. (Betulaceae) from the Paleocene of Wyoming, U.S.A. *International Journal of Plant Sciences* 157:644–655.

———. 1998. A new genus of Coryloideae (Betulaceae) from the Paleocene of North America. *International Journal of Plant Sciences* 159:522–532.

Manchester, S. R., and P. R. Crane. 1987. A new genus of Betulaceae from the Oligocene of western North America. *Botanical Gazette* 148:263–273.

Manchester, S. R., and Guo Shuang-Xing. 1996. *Palaeocarpinus* (extinct Betulaceae) from northwestern China: New evidence for Paleocene floristic continuity between Asia, North America, and Europe. *International Journal of Plant Sciences* 157:240–246.

Manchester, S. R., and L. J. Hickey. 2007. Reproductive and vegetative organs of *Browniea* gen. n. (Nyssaceae) from the Paleocene of North America. *International Journal of Plant Sciences* 168:229–249.

Manchester, S. R., D. L. Dilcher, and W. D. Tidwell. 1986. Interconnected reproductive and vegetative remains of *Populus* (Salicaceae) from the Middle Eocene Green River Formation, northeastern Utah. *American Journal of Botany* 73:156–160.

Manchester, S. R., P. R. Crane, and D. L. Dilcher. 1991. *Nordenskioldia* and *Trochodendron* (Trochodendraceae) from the Miocene of northwestern North America. *Botanical Gazette* 152:357–368.

Manchester, S. R., P. R. Crane, and L. B. Golovneva. 1999. An extinct genus with affinities to extant *Davidia* and *Camptotheca* (Cornales) from the Paleocene of North America and eastern Asia. *International Journal of Plant Sciences* 160:188–207.

Manchester, S. R., M. A. Akhmetiev, and T. M. Kodrul. 2002. Leaves and fruits of *Celtis aspera* (Newberry) comb. nov. (Celtidaceae) from the Paleocene of North America and eastern Asia. *International Journal of Plant Sciences* 163:725–736.

Martin, A.R.H. 1976. Upper Palaeocene Salviniaceae from the Woolwich/Reading beds near Cobham, Kent. *Palaeontology* 19:173–184.

Mayo, S. J., J. Bogner, and P. C. Boyce. 1997. *The genera of Araceae*. Royal Botanic Gardens, Kew, U.K.

McIver, E. E. 1992. Fossil *Fokienia* (Cupressaceae) from the Paleocene of Alberta, *Canada. Canadian Journal of Botany* 70:742–749.

———. 1994. An early *Chamaecyparis* (Cupressaceae) from the late Cretaceous of Vancouver Island, British Columbia, Canada. *Canadian Journal of Botany* 72:1787–1796.

———. 1999. Paleobotanical evidence for ecosystem disruption at the Cretaceous-Tertiary boundary from Wood Mountain, Saskatchewan, Canada. *Canadian Journal of Earth Sciences* 36:225–789.

———. 2001. Cretaceous *Widdringtonia* Endl. (Cupressaceae) from North America. *International Journal of Plant Science* 162:937–961.

———. 2002. The palaeoenvironment of *Tyrannosaurus rex* from southwestern Saskatchewan, Canada. *Canadian Journal of Earth Sciences* 39:207–221.

McIver, E. E., and K. R. Aulenback. 1994. Morphology and relationships of *Mesocyparis umbonata* sp. nov.: Fossil Cupressaceae from the late Cretaceous of Alberta. *Canadian Journal of Botany* 72:273–295.

McIver, E. E., and J. F. Basinger. 1987. *Mesocyparis borealis* gen. et sp. nov.: Fossil Cupressaceae from the early Tertiary of Saskatchewan, Canada. *Canadian Journal of Botany* 65:2338–2351.

———. 1989. The morphology and relationships of *Equisetum fluviatoides* sp. nov. from the Paleocene Ravenscrag Formation of Saskatchewan, Canada. *Canadian Journal of Botany* 67:2937–2943.

———. 1990. Fossil seed cones of *Fokienia* (Cupressaceae) from the Paleocene Ravenscrag Formation of Saskatchewan, Canada. *Canadian Journal of Botany* 68:1609–1990.

———. 1993. Flora of the Ravenscrag Formation (Paleocene), southwestern Saskatchewan, Canada. *Palaeontographica Canadiana* 10:1–167.

McIver, E. E., A. R. Sweet, and J. F. Basinger. 1991. Sixty-five-million-year-old flowers bearing pollen of the extinct triprojectate complex – a Cretaceous-Tertiary boundary survivor.

Review of Palaeobotany and Palynology 70:77–88.

Mehrotra, R. C. 1987. Some new palm fruits from the Deccan Intertrappean beds of Mandla District, Madhya Pradesh. *Geophytology* 17:204–208.

Meicenheimer, R. D. 1998. Decussate to spiral transitions in phyllotaxis. In *Symmetry in Plants*, edited by R. V. Jean and D. Barabe, 125–143. Singapore: World Scientific.

Mickel, J. T. 1981. Revision of *Anemia* subgenus *Anemiorrhiza* (Schizaeaceae). *Brittonia* 33:413–429.

Miller, C. N. Jr. 1967. Evolution of the fern genus *Osmunda*. *Contributions from the Museum of Paleontology, University of Michigan* 21:139–203.

———. 1971. Evolution of the fern family Osmundaceae based on anatomical studies. *Contributions from the Museum of Paleontology, University of Michigan* 23:105–169.

———. 1972. *Pityostrobus palmeri*, a new species of petrified conifer cones from the late Cretaceous of New Jersey. *American Journal of Botany* 59:352–358.

———. 1974. *Pityostrobus hallii*, a new species of structurally preserved conifer cones from the late Cretaceous of Maryland. *American Journal of Botany* 61:798–804.

———. 1975. Petrified cones and needle-bearing twigs of a new taxodiaceous conifer from the early Cretaceous of California. *American Journal of Botany* 62:706–713.

———. 1977. Mesozoic Conifers. *Botanical Review* 43:217–280.

———. 1982. Current status of Paleozoic and Mesozoic conifers. *Review of Palaeobotany and Palynology* 37:99–114.

———. 1988. The origin of modern conifer families. In *Origin and Evolution of Gymnosperms*, edited by C. B. Beck, 448–486. New York; Columbia University Press.

Miller, C. N. Jr., and C. A. Lapasha. 1983. Structure and affinities of *Athrotaxites berryi* Bell, an early Cretaceous conifer. *American Journal of Botany* 70:772–779.

———. 1984. Flora of the early Cretaceous Kootenai Formation in Montana, conifers. *Palaeontology Abt. B* 193:1–17.

———. 1985. Two species of *Elatocladus* from the early Cretaceous Potomac group of Virginia. *Review of Palaeobotany and Palynology* 44:183–191.

Miner, E. L. 1935. Paleobotanical examinations of Cretaceous and Tertiary coals. *American Midland Naturalist* 16:585–615.

Morley, T. 1948. On leaf arrangement in *Metasequoia glyptostroboides*. *Proceedings of the National Academy of Science, U.S.A.* 34:574–578.

Muhammad, A. F. 1986. The study and identification of five petrified woods from the Tyrrell Museum of Palaeontology Drumheller, Alberta. Service contract (unpublished).

Muller, J. 1981. Fossil pollen records of extant angiosperms. *Botanical Review* 47:1–142.

Nagalingum, N. S. 2007. *Marsileaceaephyllum*, a new genus for marsileaceous macrofossils: Leaf remains from the early Cretaceous (Albian) of southern Gondwana. *Plant Systematic Evolution* 264:41–55.

Namboodiri, K. K., and C. B. Beck. 1968a. A comparative study of the primary vascular system of conifers. I. Genera with helical phyllotaxis. *American Journal of Botany* 55:447–457.

———. 1968b. A comparative study of the primary vasculature system of conifers. II. Genera with opposite and whorled phyllotaxis. *American Journal of Botany* 55:458–463.

———. 1968c. A comparative study of the primary vasculature system of conifers. III. Stelar evolution in Gymnosperms. *American Journal of Botany* 55:464–472.

Namburdiri, E.M.V., and S. Chitaley. 1991. Fossil *Salvinia* and *Azolla* from the Deccan Intertrappean beds of India. *Review of Palaeobotany and Palynology* 69:325–336.

Nast, C. G., and I. W. Bailey. 1945. Morphology and relationships of *Trochodendron* and *Tetracentron*, II inflorescence, flower, and fruit. *Journal of the Arnold Arboretum* 26:267–276.

Nathorst, A. G. 1909. Über die Gattung *Nilssonia* Brongn. Mit besonderer Berücksichtigung schwedischer Arten. *Kungliga Svenska vetenskapsakademiens handlingar* 43:1–40.

Nayar, B. K., and S. Kaur. 1963a. Contributions to the morphology of some species of *Microlepia*. *Journal of the Indian Botanical Society* 42:225–232.

———. 1963b. Ferns of India- VIII *Microlepia* Presl. *Bulletin of the National Botanic Gardens, Lucknow, India*, No. 79.

Nishida, M., H. Nishida, and T. Ohsawa. 1991a. Structure and affinities of the petrified plants from the Cretaceous of Northern Japan and Saghalien VIII. *Parataiwania nihongii* gen. et sp. nov., a taxodiaceous cone from the Upper Cretaceous of Hokkaido. *Journal of Japanese Botany* 67:1–9.

———. 1991b. Structure and affinities of the petrified plants from the Cretaceous of northern Japan and Saghalien VI. *Yezosequoia shimanukii* gen. et sp. nov., a petrified taxodiaceous cone from Hokkaido. *Journal of Japanese Botany* 66:280–291.

Nurkowski, J. R., and R. A. Rahmani. 1984. Cretaceous fluvio-lacustrine coal-bearing sequence, Red Deer area, Alberta, Canada. In *Sedimentology of Coal and Coal-Bearing Sequences*, edited by R.A. Rahmani and R.M. Flores. International Society of Sedimentology. Special Publication 7:163–176.

Offler, C. E. 1984. Living and fossil Coniferales of Australia and New Guinea. Part 1: A study of the external morphology of the vegetative shoots of the living species. *Palaeontology Abt. B* 193:18–120.

Ogden, E. C. 1974. Anatomical patterns of some aquatic vascular plants of New York. *New York State Museum and Science Service Bulletin* 424:1–133.

Ogg, J. G. 1995. Magnetic polarity time scale of the Phanerozoic. *In: Global Earth Physics: A handbook of physical constants. AGU reference shelf* 1:240–270.

Ogura, Y. 1972. Comparative anatomy of vegetative organs of the Pteridophytes. Spezieller Teil, Band VII, Teil 3:1–502. Borntraeger, Berlin.

Ohsawa, T. 1994. Anatomy and relationships of petrified seed cones of the Cupressaceae, Taxodiaceae, and Sciadopityaceae. *Journal of Plant Research* 107:503–512.

Ohsawa, T., H. Nishida, and M. Nishida. 1992a. Structure and affinities of the petrified plants from the Cretaceous of Northern Japan and Saghalien XI. A cupressoid seed cone from the Upper Cretaceous of Hokkaido. *Botanical Magazine Tokyo* 105:125–133.

Ohsawa, T., M. Nishida, and H. Nishida. 1992b. Structure and affinities of the petrified plants from the Cretaceous of Northern Japan and Saghalien XII. *Obirastrobus* gen. nov., petrified pinaceous cones from the Upper Cretaceous of Hokkaido. *Botanical Magazine Tokyo* 105:461–484.

———. 1992c. Structure and affinities of the petrified plants from the Cretaceous of northern Japan and Saghalien X. Two *Sequoia*-like cones from the Upper Cretaceous of Hokkaido. *Journal of Japanese Botany* 67:72–82.

Ohsawa, T., H. Nishida, and M. Nishida. 1993. Structure and affinities of the petrified plants from the Cretaceous of northern Japan and Saghalien XIII. *Yubaristrobus* gen. nov., a new taxodiaceous cone from the Upper Cretaceous of Hokkaido. *Journal of Plant Research* 106:1–9.

Oladele, F. A. 1983a. Scanning electron microscope study of stomatal-complex configuration in Cupressaceae. *Canadian Journal of Botany* 61:1232–1240.

———. 1983b. Inner surface sculpture patterns of cuticles in Cupressaceae. *Canadian Journal of Botany* 61:1222–1231.

Page, V. M. 1967. Angiosperm wood from the Upper Cretaceous of central California: Part I. *American Journal of Botany* 54:510–514.

———. 1968. Angiosperm wood from the Upper Cretaceous of central California: Part II. *American Journal of Botany* 55:168–172.

———. 1970. Angiosperm wood from the Upper Cretaceous of central California: Part III. *American Journal of Botany* 57:1139–1144.

Pant, D. D. 1962. *Studies in Gymnospermous Plants: Cycads.* Allahabad: Central Book Depot.

———. 1987. The fossil history and phylogeny of the Cycadales. *Geophytology* 17:125–162.

Parker, L. R. 1976. The Paleoecology of the fluvial coal-forming swamps and associated floodplain environments in the Blackhawk Formation (Upper Cretaceous) of Central Utah. *Geological Studies, Brigham Young University* 22:99–116.

Patton Jr., W. W., and E. J. Moll-Stalcup. 2000. Geologic map of the Nulato Quadrangle, west-central Alaska. *United States Geological Survey, Map* I-2677 and pamphlet.

Penfound, W. T. 1952. Southern swamps and marshes. *Botanical Review* 18:413–446.

Penhallow, D. P. 1908. Report on a collection of fossil woods from the Cretaceous of Alberta. *Ottawa Naturalist* July: 82–88.

Penny, J. S. 1947. Studies on the conifers of the Magothy flora. *American Journal of Botany* 34:281–296.

Peppe, D. J., M. J. Erickson, and L. J. Hickey. 2007. Fossil leaf species from the Fox Hills Formation (Upper Cretaceous: North Dakota, USA) and their paleogeographic significance. *Journal of Paleontology* 81:550–567.

Peters, M. D. 1985. A taxonomic analysis of a middle Cretaceous megafossil plant assemblage from Queensland, Australia. PhD thesis, University of Adelaide, Australia.

Peters, M. D, and D. C. Christophel. 1978. *Austrosequoia wintonensis*, a new taxodiaceous cone from Queensland, Australia. *Canadian Journal of Botany* 56:3119–3128.

Poole, I., and D. Cantrill. 2001. Fossil woods from Williams Point Beds, Livingstone Island, Antarctica: A late Cretaceous southern high latitude flora. *Palaeontology* 44:1081–1112.

Poole, I., H. G. Richter, and J. E. Francis. 2000. Evidence for Gondwanan origins for *Sassafras*

(Lauraceae?) late Cretaceous fossil wood of Antarctica. *IAWA Journal* 21:463–475.

Poole, J. P. 1923. Comparative anatomy of leaf of Cycads, with reference to Cycadofilicales. *Botanical Gazette* 76:203–214.

Postnikoff, D. 1997. The addition of *Taiwania* sp. to the fossil conifers of the Tertiary fossil forest of Axel Heiberg Island, Canadian Arctic Archipelago. Unpublished honours thesis, University of Saskatchewan.

Price, R. A., and J. M. Lowenstein. 1989. An immunological comparison of the Sciadopityaceae, Taxodiaceae, and Cupressaceae. *Systematic Botany* 14:141–149.

Pryer, K. M. 1999. Phylogeny of marsileaceous ferns and relationships of the fossil *Hydropteris pinnata* reconsidered. *International Journal of Plant Sciences* 160:931–954.

Pryer, K. M., E. Schuettpelz, P. G. Wolf, H. Schneider, A. R. Smith, and R. Cranfill. 2004. Phylogeny and evolution of ferns (monilophytes) with a focus on the early leptosporangiate divergences. *American Journal of Botany* [Special invited paper] 91:1582–1598.

Pryer, K. M., A. R. Smith, and J. E. Skog. 1995. Phylogenetic relationships of extant ferns based on evidence from morphology and rbcL sequences. *American Fern Journal* 85:205–282.

Puri, V., and M. L. Garg. 1953. A contribution to the anatomy of the sporocarp of *Marsilea minuta* L. with a discussion of the nature of sporocarp in the Marsileaceae. *Phytomorphology* 3:190–209.

Rahmani, R. A., and L. V. Hills. 1982. Facies relationships and paleoenvironments of a late Cretaceous tide-dominated delta, Drumheller, Alberta. Field Trip Guidebook No. 6, Canadian Society of Petroleum Geologists, Calgary, Alberta.

Ramanujam, C.G.K. 1972. Fossil coniferous woods from the Oldman Formation (Upper Cretaceous) of Alberta. *Canadian Journal of Botany* 50:595–602.

Ramanujam, C.G.K., and W. N. Stewart. 1969a. Fossil woods of Taxodiaceae from the Edmonton Formation (Upper Cretaceous) of Alberta. *Canadian Journal of Botany* 47:115–124.

———. 1969b. Taxodiaceous bark from the Upper Cretaceous of Alberta. *American Journal of Botany* 56:101–107.

Rao, H. S. 1935. The structure and life history of *Azolla pinnata* R. Brown with remarks on the fossil history of the Hydropteridae. *Proceedings of the Indian Academy of Science* 2:175–200.

Reed, C. F. 1947. The phylogeny and ontogeny of the Pteropsida: I. Schizaeales. *Boletim Sociedade Broteriana* 21:71–197.

Richards, F. J. 1951. Phyllotaxis: Its quantitative expression and relation to growth in the apex. *Philosophical Transactions of the Royal Society of London; Series B* 235:509–564.

Robison, C. R. 1977. *Pinus triphylla* and *Pinus quinquefolia* from the Upper Cretaceous of Massachusetts. *American Journal of Botany* 64:726–732.

Roig, F. A. 1992. Comparative wood anatomy of southern South American Cupressaceae. *IAWA Bulletin*, N.S. 13:151–162.

Rothwell, G. W., and J. F. Basinger. 1979. *Metasequoia milleri* n. sp., anatomically preserved pollen cones from the middle Eocene (Allenby Formation) of British Columbia. *Canadian Journal of Botany* 57:958–970.

Rothwell, G. W., and B. Holt. 1997. Fossils and phenology in the evolution of *Ginkgo biloba*. In *Ginkgo biloba – A Global Treasure from Biology to Medicine*, edited by T. Hori, R. W. Ridge, W. Tulecke, et al., 223–230. Tokyo: Springer.

Rothwell, G. W., and R. A. Stockey. 1991. *Onoclea sensibilis* in the Paleocene of North America, a dramatic example of structural and ecological stasis. *Review of Palaeobotany and Palynology* 70:113–124.

———. 1994. The role of *Hydropteris pinnata* gen. et sp. nov. in reconstructing the phylogeny

REFERENCES AND SUGGESTED READINGS

of heterosporous ferns. *American Journal of Botany* 81:479–492.

Rouse, G. E. 1967. A late Cretaceous plant assemblage from east-central British Columbia: I, fossil leaves. *Canadian Journal of Earth Sciences* 4:1185–1197.

Roy, S. K. 1972. Fossil wood of Taxaceae from the McMurray Formation (Lower Cretaceous) of Alberta, Canada. *Canadian Journal of Botany* 50:349–352.

Saiki, K. 1991. A cladistics analysis of the Taxodiaceae based on their female cones. PhD thesis, Geological Institute, University of Tokyo.

Saiki, K., and T. Kimura. 1993. Permineralized taxodiaceous seed cones from the Upper Cretaceous of Hokkaido, Japan. *Review of Palaeobotany and Palynology* 76:83–96.

Samylina, V. A. 1962. On the Cretaceous flora of the Arkagala Coal Basin. *Doklady Academia Nauka SSSR* 147:1157–1159 (in Russian).

———. 1964. Cretaceous flora of the Arkagala Coal Basin. *Doklady of the Academy of Sciences of the U.S.S.R., Earth Science Section* 147:113–115 (in Russian).

Sander, M. P., and C. T. Gee. 1990. Fossil charcoal: Techniques and applications. *Review of Palaeobotany and Palynology* 63:269–279.

Saunders, R.M.K., and K. Fowler. 1992. A morphopogical taxonomic revision of *Azolla* Lam. Section Rhizosperma (Mey.) Mett. (Azollaceae). *Botanical Journal of the Linnean Society* 109:329–357.

———. 1993. The supraspecific taxonomy and evolution of the fern genus *Azolla* (Azollaceae). *Plant Systematic Evolution* 184:175–193.

Schwarz, O., and H. Weide. 1962. Systematische Revision der Gattung *Sequoia* Endl. *Feddes Repertorium.* 16:159–192.

Scott, E. S., P. L. Williams, L. C. Craig, E. S. Barghoorn, L. J. Hickey, and H. D. MacGinitie. 1972. "Pre-Cretaceous" angiosperms from Utah: Evidence for Tertiary age of the Palm woods and roots. *American Journal of Botany* 59:886–896.

Scott, R. A., E. S. Barghoorn, and U. Prakash. 1962. Wood of *Ginkgo* in the Tertiary of western North America. *American Journal of Botany* 49:1095–1101.

Serbet, R. 1997. Morphologically and anatomically preserved fossil plants from Alberta, Canada: A flora that supported the dinosaur fauna during the Upper Cretaceous (Maastrichtian). PhD thesis, Faculty of the College of Arts, Ohio University, Athens, OH, 1–305.

Serbet, R., and G. W. Rothwell. 1999. *Osmunda cinnamomea* (Osmundaceae) in the Upper Cretaceous of western North America: Additional evidence for exceptional species longevity among filicalean ferns. *International Journal of Plant Sciences* 160:425–433.

———. 2003. Anatomically preserved ferns from the late Cretaceous of western North America: Dennstaedtiaceae. *International Journal of Plant Sciences* 64:1041–1051.

———. 2006. Anatomically preserved ferns from the late Cretaceous of western North America. II. Blechnaceae/ Dryopteridaceae. *International Journal of Plant Sciences* 167:703–709.

Serbet, R., and R. A. Stockey. 1991. Taxodiaceous pollen cones from the Upper Cretaceous (Horseshoe Canyon Formation) of Drumheller Alberta, Canada. *Review of Paleobotany and Palynology* 70:67–76.

Seward, A. C. 1969. *Fossil Plants: A Text-Book for Students of Botany and Geology.* New York: Hafner.

Shah, J. J., and E. C. Raju. 1975. General morphology, growth and branching behaviour of the rhizomes of Ginger, Tumeric and Mango Ginger. *New Botanist* 2:59–69.

Shaparenko, K. K. 1956. Istoriya sal'vinii (History of Salvinias). *Paleobotanika, Akad. Nauk SSSR, Bot. Inst.,* Tr. S. 8, Issue 2:7–43.

Sharma, B. D. 1970a. On the structure of the seeds of *Williamsonia* collected from the middle Jurassic rocks of Amarjola in the Rajmahal Hills, India. *Annals of Botany* 34:1071–1078.

——. 1970b. On the vascular organization of the receptacles of seed-bearing Williamsonias from the middle Jurassic rocks of Amarjola in the Rajmahal Hills, India. *Annals of Botany* 34:1063–1070.

——. 1974. Ovule ontogeny in *Williamsonia* Carr. *Palaeontographica Abt. B* 148:137–143.

——. 1977. Indian Williamsonias – an illustrated review. *Acta Palaeobotanica* 18:19–29.

Shimakura, M. 1937. Studies in fossil woods from Japan and adjacent lands. Contributions II. The Cretaceous woods from Japan, Saghalien and Manchoukuo. *Science Report Tohoku Imperial University, Ser. 2*, 19:1–73.

Skog, J. E. 1982. *Pelletixia amelguita*, a new species of fossil fern in the Potomac Group (Lower Cretaceous). *American Fern Journal* 72:115–121.

——. 1992. The Lower Cretaceous ferns in the genus *Anemia* (Schizaeaceae), Potomac group of Virginia, and relationships within the genus. *Review of Palaeobotany and Palynology* 70:279–295.

Skog, J. E., and D. L. Dilcher. 1992. A new species of *Marsilea* from the Dakota Formation in central Kansas. *American Journal of Botany* 79:982–988.

Smith, A. G., and J. C. Briden. 1977. *Mesozoic and Cenozoic Paleocontinental Maps.* Cambridge: Cambridge University Press.

Smith, A. R., K. M. Pryer, E. Schuettpelz, P. Korall, H. Schneider, and P. G. Wolf. 2006. A classification for extant ferns. *Taxon* 55:705–731.

Snead, R. G. 1969. Microflora diagnosis of the Cretaceous- Tertiary boundary, central Alberta. Research Council of Alberta. Bulletin 25.

Spicer, R. A., and A. G. Greer. 1986. Plant taphonomy in fluvial and lacustrine systems. In *Land Plant Notes for a Short Course*, ed. T. Broadhead, 10–26. Department of Geological Sciences, University of Tennessee, Knoxville.

Spicer, R. A., and A. B. Herman. 1995. *Nilssoniocladus* in the Cretaceous Arctic: New species and biological insights. *Review of Palaeobotany and Palynology* 92:229–243.

Srivastava, S. K. 1970. Pollen biostratigraphy and paleoecology of the Edmonton Formation (Maastrichtian), Alberta, Canada. *Palaeogeography, Palaeoclimatology, Palaeoecology* 7:221–276.

Stafford, P. J. 1995. The northwest European pollen flora, 52, Marsileaceae. *Review of Palaeobotany and Palynology* 88:3–24.

——. 2003a. The northwest European pollen flora, 58: Azollaceae. *Review of Palaeobotany and Palynology* 123:9–17.

——. 2003b. The northwest European pollen flora, 59: Salviniaceae. *Review of Palaeobotany and Palynology* 123:19–25.

Stebbins, G. L. Jr. 1948. The chromosomes and relationships of *Metasequoia* and *Sequoia*. *Science* 108:95–98.

Sterling, C. 1945a. Growth and vascular development in the shoot apex of *Sequoia sempervirens* (Lamb.) Endl. II. Vascular development in relationship to phyllotaxis. *American Journal of Botany* 32:380–386.

——. 1945b. Growth and vascular development in the shoot apex of *Sequoia sempervirens* (Lamb.) Endl. I. Structure and growth of the shoot apex. *American Journal of Botany* 32:118–136.

——. 1949. Some features in the morphology of *Metasequoia*. *American Journal of Botany* 36:461–471.

Stevens, P. F., (2001 onwards), Angiosperm Phylogeny website. Version 7, May 2006 [and more or less continuously updated since]; http://www.mobot.org/MOBOT/research/APweb/.

Stewart, W. N. 1987. Paleobotany and the evolution of plants. Cambridge University Press.

Stewart, W. N., and G. W. Rothwell. 1993. *Paleobotany and the Evolution of Plants*, 2nd ed. Cambridge: Cambridge University Press.

Stockey, R. A. 1994. Mesozoic Araucariaceae: Morphology and systematic relationships. *Journal of Plant Research* 107:493–502.

Stockey, R. A., and H. Ko. 1986. Cuticle micromorphology of *Araucaria* De Jussieu. *Botanical Gazette* 147:508–548.

Stockey, R. A., and M. Nishida. 1986. *Pinus haboroensis* sp. nov. and the affinities of permineralized leaves from the Upper Cretaceous of Japan. *Canadian Journal of Botany* 64:1856–1866.

Stockey, R. A., and G. W. Rothwell. 1997. The aquatic angiosperm *Trapago angulata* from the Upper Cretaceous (Maastrichtian) St. Mary River Formation of southern Alberta. *International Journal of Plant Sciences* 158:83–94.

Stockey, R. A., and Y. Ueda. 1986. Permineralized pinaceous leaves from the Upper Cretaceous of Hokkaido. *American Journal of Botany* 73:1157–1162.

Stockey, R. A., G. L. Hoffman, and G. W. Rothwell. 1997. The fossil monocot *Limnobiophyllum scutatum*: Resolving the phylogeny of Lemnaceae. *American Journal of Botany* 84:355–368.

Stockey, R. A., G. W. Rothwell, and A. Falder. 2001. Diversity among taxodioid conifers: *Metasequoia foxii* sp. nov. from the Paleocene of central Alberta, Canada. *International Journal of Plant Sciences* 162:221–234.

Stopes, M. C. 1910. The internal anatomy of 'Nilssonia orientalis'. *Annals of Botany* 24:389–393.

———. 1918. New Bennettitean cones from the British Cretaceous. *Philosophical Transactions of the Royal Society of London; Series B* 208:389–440.

Stopes, M. C., and E. M. Kershaw. 1910. The anatomy of Cretaceous Pine leaves. *Annals of Botany* 24:395–402.

Straight, W. H., and D. A. Eberth. 2002. Testing the utility of vertebrate remains in recognizing patterns in fluvial deposits: An example from the Lower Horseshoe Canyon Formation, Alberta. *Palaios* 17:472–490.

Strasburger, E. 1873. Ueber *Azolla*. Friedrich von Zezschwitz (vormals Fr. Eugen Köhlers Botanischer Verlag). 1–86.

Sutherland, M. 1934. A microscopical study of the structures of the leaves of the genus *Pinus*. *Transactions, Proceedings of the Royal Society of New Zealand, Wellington* 63:517–569.

Suzuki, Y. 1995. On the structure and affinities of two new conifers and a new fungus form the Upper Cretaceous of Hokkaido (Yezo). *Botanical Magazine* 24:181–196.

Svenson, H. K. 1944. The new world species of *Azolla*. *American Fern Journal* 34:69–84.

Sweet, A. R. 1972. The taxonomy, evolution, and stratigraphic value of *Azolla* and *Azollopsis* in the Upper Cretaceous and early Tertiary. Unpublished PhD thesis, University of Alberta.

———. 1974. A detailed study of the genus *Azollopsis*. *Canadian Journal of Botany* 52:1625–1642.

Sweet, A. R., and A. Chandrasekharam. 1973. Vegetative remains of *Azolla schopfii* Dijkstra from Genesee, Alberta. *Canadian Journal of Botany* 51:1491–1496.

Sweet, A. R., and L. V. Hills. 1971. A study of *Azolla pinnata* R. Brown. *American Fern Journal* 61:1–13.

Sweet, A. R., and L. V. Hills. 1976. Early Tertiary species of *Azolla* subg. *Azolla* sect. *Kremastospora* from western and arctic Canada. *Canadian Journal of Botany* 54:334–351.

Takahashi, M., P. Crane, and H. Ando. 1999a. Fossil flowers and associated plant fossils from the Kamikitaba locality (Ashizawa Formation, Futaba Group, Lower Coniacian, Upper Cretaceous) of northeast Japan. *Journal of Plant Research* 112:187–206.

———. 1999b. *Esgueiria futabensis* sp. nov., a new angiosperm flower from the Upper Cretaceous (Lower Coniacian) of northeastern Honshu, Japan. *Palaeontology Research* 3:81–87.

———. 2001. Fossil megaspores of Marsileales and Selaginellales from the Upper Coniacian to Lower Santonian (Upper Cretaceous) of the Tamagawa Formation (Kiju group) in northeastern Japan. *International Journal of Plant Sciences* 162:431–439.

Takaso, T., and P. B. Tomlinson. 1989. Aspects of cone and ovule ontogeny in *Cryptomeria* (Taxodiaceae). *American Journal of Botany* 76:692–705.

———. 1990. Cone and ovule ontogeny in *Taxodium* and *Glyptostrobus* (Taxodiaceae-Coniferales). *American Journal of Botany* 77:1209–1221.

———. 1992. Seed cone and ovule ontogeny in *Metasequoia*, *Sequoia* and *Sequoiadendron* (Taxodiaceae-Coniferales). *Botanical Journal of the Linnean Society* 109:15–37.

Taylor, T. N. 1981. *Paleobotany: An Introduction to Fossil Plant Biology*, ed. J.E. Vastyan and J.S. Amar. New York: McGraw-Hill.

Taylor, T. N., and E. L. Taylor. 1993. *The Biology and Evolution of Fossil Plants*. Englewood Cliffs, NJ: Prentice-Hall.

Thayn, G. F., W. D. Tidwell, and W. L. Stokes. 1985. Flora of the Lower Cretaceous Cedar Mountain Formation of Utah and Colorado. Part III: *Icacinoxylon pittiense* N. Sp. *American Journal of Botany* 72:175–180.

Thomas, B. A. 1987. The use of in-situ spores for defining species of dispersed spores. *Review of Palaeobotany and Palynology* 51:227–233.

Thornley, J. H. M. 1975a. Phyllotaxis I: A mechanistic model. *Annals of Botany* 39:491–507.

———. 1975b. Phyllotaxis II: A description in terms of intersecting logarithmic spirals. *Annals of Botany* 39:509–524.

Tidwell, W. D. 1998. *Common Fossil Plants of Western North America*, 2nd ed. Washington: Smithsonian Institution Press.

Tidwell, W. D., and E.M.V. Nambudiri. 1989. *Tomlinsonia thomassonii*, gen. et sp. nov., a permineralized grass from the Upper Miocene Ricardo Formation, California. *Review of Palaeobotany and Palynology* 60:165–177.

Tiffney, B. H. 1977. Fossil Angiosperm fruits and seeds. *Journal of Seed Technology* 2:54–71.

Tomlinson, P. B. 1956. Studies in the systematic anatomy of the Zingiberaceae. *Botanical Journal of the Linnean Society* 55:547–592.

Tomlinson, P. B., and M. H. Zimmermann. 1966. Anatomy of the palm *Rhapis excelsa* – II. rhizome. *Journal of the Arnold Arboretum* 47:248–261.

Tomlinson, P. B., T. Takaso, and E. K. Cameron. 1993. Cone development in *Libocedrus* (Cupressaceae)-phenological and morphological aspects. *American Journal of Botany* 80:649–659.

Tozer, E. T. 1956. Uppermost Cretaceous and Paleocene non-marine molluscan faunas of western Alberta. Geological Survey of Canada Memoir 280, no. 2521: 1–125.

Tralau, H. 1968. Evolutionary trends in the genus *Ginkgo*. *Lethaia* 1:63–101.

Tryon, R. M., and A. F. Tryon. 1982. *Ferns and Allied Plants: With Special Reference to Tropical America*. New York: Springer.

Tryon, A. F., and B. Lugardon. 1991. *Spores of the Pteridophyta*. New York: Springer.

Ueda, Y., and M. Nishida. 1982. On the petrified pine leaves from the Upper Cretaceous of Hokkaido. *Journal of Japanese Botany* 57:133–145.

Upchurch, G. R. Jr., and J. A. Wolfe. 1993. Cretaceous vegetation of the western interior and adjacent regions of North America. In *Evolution of the Western Interior Basin*, ed. E. G. Kauffman and W. G. E. Caldwell, 243–281. Geological Association of Canada Special Paper 39.

Vinnersten, A., and K. Bremer. 2001. Age and biogeography of major clades in Liliales. *American Journal of Botany* 88:1695–1703.

Wagner, W. H. 1996. Deciduous vs. marcescent ferns and fern allies. *Fiddlehead Forum Bulletin of the American Fern Society* 23:32–33.

Wan, Z. 1996. The Lower Cretaceous flora of the Gates Formation from western Canada. PhD thesis, University of Saskatchewan.

Wang, Chi-Wu. 1961. The forests of China. Maria Moore Cabot Foundation Publication Series 5.

Watson, J., and C. A. Sincock. 1992. Bennettitales of the English Wealden. Monograph Palaeontology Society, London 145.

Weber, R. 1973. *Salvinia coahuilensis* nov. sp. del Cretacico Superior de Mexico. *Ameghiniana* 10:173–190.

———. 1976. *Dorfiella auriculata* f. gen. nov., sp. nov. un genero nuevo de helechos acuaticos del Cretacico Superior de Mexico. *Del Boletin de la Asociacion Latinoamericana de Paleobotanica y Palinologia* 3:1–13 (in Spanish).

Wheeler, E. A. and T. M. Lehman. 2000. Late Cretaceous woody dicots from the Aguja and Javelina formations, Big Bend National Park, Texas, USA. *IAWA Journal* 21:83–120.

———. 2005. Upper Cretaceous-Paleocene conifers woods from Big Bend National Park, Texas. *Palaeogeography, Palaeoclimatology, Palaeoecology* 226:233–258.

Wheeler, E. A., T. M. Lehman, and P. E. Gasson. 1994. *Javelinoxylon*, an Upper Cretaceous dicotyledonous tree from Big Bend National Park, Texas, with presumed malvalean affinities. *American Journal of Botany* 81:703–710.

White, M. E. 1990. *The Greening of Gondwana*. Balgowlah, NSW: Reed Books.

Wolfe, J. A. 1987. Late Cretaceous-Cenozoic history of deciduousness and the terminal Cretaceous event. *Paleobiology* 13:215–226.

Wolfe, J. A., and G. R. Upchurch Jr. 1987. North American nonmarine climates and vegetation during the late Cretaceous. *Palaeogeography, Palaeoclimatology, Palaeoecology* 61:33–77.

Wright, A. H., and A. A. Wright. 1932. The habitats and composition of the vegetation of Okefinokee Swamp, Georgia. *Ecological Monographs* 2:109–232.

Wu, X.-C., D. B. Brinkman, and A. P. Russell. 1996. A new alligator from the Upper Cretaceous of Canada and the relationships of early eusuchians. *Palaeontology* 39:351–375.

Xi, Y. Z., and F. H. Wang. 1989. Pollen exine ultrastructure of living Chinese gymnosperms. *Cathaya* 1:119–142.

Yamada, T., and M. Kato. 2002. *Regnellites nagashimae* gen. et sp nov., the oldest macrofossil of Marsileaceae, from the Upper Jurassic to Lower Cretaceous of western Japan. *International Journal of Plant Science* 163:715–723.

Yao, X., T. N. Taylor, and E. L. Taylor. 1997. A taxodiaceous seed cone from the Triassic of Antarctica. *American Journal of Botany* 84:343–354.

Yao, X., Z. Zhou, and B. Zhang. 1998. Reconstruction of the Jurassic conifer *Sewardiodendron laxum* (Taxodiaceae). *American Journal of Botany* 85:1289–1300.

Young, J. A., and C. G. Young. 1992. *Seeds of Woody Plants in North America*. Portland, OR: Dioscorides Press.

Zavada, M. S. 2007. The identification of fossil angiosperm pollen and its bearing on the time and space of the origin of angiosperms. *Plant Systematic Evolution* 263:117–134.

Zhang, Z. C. 1985. Main stages of Cretaceous angiosperm succession in north part of northeast China. *Acta Paleontologica Sinica* 24:453–460.

Zhou, Z., W. L. Crepet, and K.C. Nixon. 2001. The earliest fossil evidence of the Hamamelidaceae: Late Cretaceous (Turonian) inflorescences and fruits of Altingioideae. *American Journal of Botany* 88:753–766.

Zhiyan, Z. 1987. *Elatides harrisii*, sp. nov., from the Lower Cretaceous of Liaoning, China. *Review of Palaeobotany and Palynology* 51:189–204.

INDEX TO THE IDENTIFICATION GUIDE TO THE FOSSIL PLANTS OF HORSESHOE CANYON FORMATION

Introductory Note: Family names are in block letters. Scientific names (Genus and species) are in italics. Page numbers in italics are noted for figures. Page numbers in bold indicate topic headings.